Hal Williams



# Mechanical Refrigeration

Elibron Classics
www.elibron.com

Elibron Classics series.



ISBN 0-543-76742-6 (paperback)
ISBN 0-543-76741-8 (hardcover)

This Elibron Classics Replica Edition is an unabridged facsimile of the edition published in 1903 by Whittaker, London; New York.

# MECHANICAL REFRIGERATION.

# MECHANICAL REFRIGERATION

BEING

A PRACTICAL INTRODUCTION TO THE STUDY OF COLD STORAGE, ICE-MAKING, AND OTHER PURPOSES TO WHICH REFRIGERATION IS BEING APPLIED.

BY

HAL WILLIAMS,

A.M.I.MECH.E., A.M.I.E.E.

WITH 115 ILLUSTRATIONS.

WHITTAKER & CO.,

2 WHITE HART ST., PATERNOSTER SQ., LONDON,

AND 66 FIFTH AVENUE, NEW YORK.

1903.

# PREFACE.

WITH the exception of M. Ledoux' work on the properties of the various refrigerating agents published in 1879, Mechanical Refrigeration for some years possessed no literature of its own. More recently several works dealing with the subject have been published. Prominent amongst these is Professor J. E. Siebel's valuable book, which, though consisting largely of tables and formulæ, concisely covers the whole field. The other works, for the most part, confine themselves to a description of the various machines and processes by which Refrigeration is obtained. Up to the present no work has been published which deals comprehensively with the important questions of Insulation, Ice-making, and Cold Storage viewed more particularly from the standpoint of the owner and user. Consequently, when I was asked by the publishers to prepare this volume, I decided as far as possible to write from this point of view and to confine myself to the practical issues rather than to the purely technical. It may perhaps be thought that having this object in view I have devoted too much space to theory, but it seems to me that the very fact that Mechanical Refrigeration is a new and growing industry makes it particularly important that those people who contemplate employing it should know something of the theory which governs the practice. It is always better to be sure than sorry, and if the limits outside which it is neither wise nor profitable to venture are clearly understood, practice within those limits may be varied to any degree that the circumstances of any particular case may call for.

Considerable misapprehension has existed regarding the behaviour of $CO_2$ at temperatures near its critical point, and I have, with the valuable assistance of Mr Alfred W. Porter, B.Sc. (Fellow of, and Assistant-Professor of Physics in, University College), endeavoured to throw some light on the subject, making free use of the researches of Dr Mollier and the entropy diagram for the purpose. It may also be thought that I have passed over the Cold-Air, Sulphuric-Dioxide, and Ammonia-Absorption Machines too lightly; this, however, has been due, not to a want of appreciation on my part of their respective merits, but to the question of space. When writing on a comprehensive subject, one must either say something of little value about a lot of things or endeavour to say something of value about a few things. Rightly or wrongly, I believe that the subjects I have dealt with are the most important, and so I have confined myself to them.

The chapter on Auxiliary Plant will, I hope, be found useful. A cold store or ice factory is a carefully thought out combination of machinery and buildings, and very frequently I have found that though considerable attention has been paid to the efficiency of the refrigerating machinery, the auxiliary machinery and appliances have to a great extent been left to look after themselves. Economy in running depends very largely upon the efficiency of the auxiliary machinery, and it is just as worthy of consideration as anything else.

I wish it to be clearly understood, however, that I am not recommending the machinery and appliances, which I have described in this chapter and elsewhere, to the exclusion of any other make. They are only to be regarded as good examples of their respective classes.

Part of Chapter XIV. has appeared in *Cold Storage and Ice Trades Review*; a portion of Chapter X. is reprinted from the *Proceedings of the Cold Storage and Ice Association*, and has appeared in *The Engineer* and other papers. A few of the blocks used in Chapter VII. have been borrowed from *Central Station Electricity Supply*, by Messrs Gay & Yeaman, to whom my thanks are due.

In conclusion, I desire to thank those friends who have so kindly given me assistance, as well as those firms who have so courteously given me drawings of their specialities. I also desire to record my indebtedness to the Technical Press, whose columns always contain so much valuable information.

HAL WILLIAMS.

60 QUEEN VICTORIA STREET,
LONDON, E.C., *April* 1903.

# CONTENTS.

## INTRODUCTION.

THE GENERAL PRINCIPLES.

## CHAPTER I.

THE LAWS OF HEAT, FLUIDS, LIQUIDS, GASES, AND VAPOURS.



## CHAPTER II.

THERMO-DYNAMICS AND ENTROPY.

## CHAPTER III.

### HISTORICAL.



## CHAPTER IV.

### THE PROPERTIES OF AMMONIA AND $CO_2$.



## CHAPTER V.

### TYPES OF MACHINES.



## CHAPTER VI.

### THE REFRIGERATING PLANT.

## CHAPTER X.

### DISTILLED WATER.



## CHAPTER XI.

### ICE-MAKING.



## CHAPTER XII.

### COLD STORAGE.



## CHAPTER XIII.

### ARTICLES IN COLD STORAGE.

## CHAPTER XIV.

### OTHER APPLICATIONS.

*Refrigeration in the Brewery.*



## TABLES.

# MECHANICAL REFRIGERATION.

## INTRODUCTION.

### THE GENERAL PRINCIPLES.

THOSE people who have been sufficiently interested to give any thought to the subject are familiar with the phenomenon of the conversion of heat into work, as exemplified in the steam-engine and in other heat-engines. The operation of the refrigerating machine may best be described as the direct opposite of that process, namely, the conversion of work into heat in such a way —paradoxical as it may seem—as to result in the production of cold. The steam-engine had been at work for many years before the laws governing its action were guessed at or understood, and in the same way most arts and sciences, as practised to-day, have been no invention, but have originated in the dim past, have been the direct result of the common need of things, and have developed by the exercise of ordinary common-sense and instinct, by rule of thumb, and by trial and error. The science of refrigeration, on the other hand, shares with that of electricity the distinction of being the direct outcome of scientific research. The whole development of each has been guided in a remarkable manner by theory derived from the study of scientific facts; and, as in 1831 the study of the laws of magnetism enabled Faraday to discover the possibility of turning mechanical into electrical energy by means of electro-magnetic induction, so a study of the laws of thermo-dynamics enabled men like Gorrie, Harrison, Perkins, Coleman, Linde, and Windhausen, to mention only a few, to discover, invent, and work apparatus for the removal of heat, or, in other words, the production of cold.

One of the most remarkable, and to the uninitiated one of the most puzzling, things about refrigeration is the fact that in order to produce cold, heat must be expended. The explanation is really very simple. The first law of thermo-dynamics tells us that both heat and work, or mechanical energy, are mutually convertible; and the second—and this is the important point—that heat cannot be raised from a lower to a higher level of temperature (that is to say, cannot pass from a cold body to a hot body) without the expenditure of some external energy. Owing to a confusion of the meaning of the two terms "heat" and "cold" and of the origin of heat, cold is often in error spoken of as if it were some substance complete in itself. Cold is merely an absence of heat, and people should not speak of "allowing cold to escape" when they mean "allowing heat to enter." The production of a state of cold consists entirely in depriving a body of its heat; it is consequently necessary, in order to conform with the second law of thermo-dynamics, to expend external energy, and, as heat is the source of all energy, and the steam-engine is one of the most economical means of turning heat into work, it will be found that the great bulk of modern refrigerating machines consist of a steam-engine working a heat-pump. The remainder consist of machines which employ the direct action of heat, and do not make use of mechanical work at all.

As water is always trying to find its own level, so heat is always endeavouring to balance itself, and when once a substance is cooled or refrigerated it will tend to absorb heat from some warmer substance until its temperature is again that of its surroundings. The term refrigeration suggests that the reduction in temperature of a body should be maintained as well as originated, and to effect this it is necessary, by means of some medium which is capable of absorbing heat, to keep on pumping the heat out of the body to be kept cool as fast as it flows in. Pumping heat is perhaps an unusual expression, but it is the principle underlying refrigeration. The heat absorbed from the cold body is raised by a compressor or pump to a high level of temperature—that is to say, to a temperature higher than the normal temperature of some substance that will, while in its natural condition, receive heat from it. In practice this substance is always water, which is circulated around pipes or over coils containing the heated medium. By this means an exchange of temperature takes place, the heat tends to equalize itself, and the water becomes warmed while the medium becomes cooled.

In a heat-engine, heat at a high temperature is taken from the coal or other source, is put into the engine, and, after doing a certain amount of work which appears as mechanical energy, is exhausted or rejected at a low temperature. In a steam-engine, for instance, high pressure or hot steam is admitted behind the piston and forces it along the cylinder; as the piston moves the steam behind it expands, and the temperature drops. The higher the high temperature ($T_1$) at which heat is taken in, and the lower the low temperature ($T_2$) at which it is exhausted, the more efficient the engine. Expressed as an equation, the efficiency $E$ of the heat-engine is

$$E = \frac{T_1 - T_2}{T_1},$$

while the efficiency $E_1$ of the refrigerating machine is exactly opposite, or

$$E_1 = \frac{T_2}{T_1 - T_2},$$

and the smaller the difference between $T_1$, the high temperature at which the heat is rejected, and $T_2$, the low temperature at which it is taken in, or, in other words, the smaller the "head" against which the heat has to be pumped in being raised from the lower to the higher level of temperature, the more efficient will the refrigerating machine be.

As a consequence of this, in a refrigerating machine driven by a steam-engine there is, in order to get the most perfect combination possible; at the one end a steam cylinder with the difference between $T_1$ and $T_2$ as *big* as possible, and at the other a compressor with the difference between them as *small* as possible.

Refrigerating machines may be roughly divided into two classes:—(1) machines which use air, and (2) those which use some chemical agent having a low boiling-point, and which produce their refrigerating effect by alternately condensing and vaporizing this agent. Of this latter class there are two sub-divisions:—(*a*) those which use heat direct to produce cold, or the absorption machines; and (*b*) those which produce cold by the expenditure of mechanical energy in working a heat-pump, or the compression machines.

Cold-air machines are also of two classes:—(1) those in which the same air is used over and over again without coming actually into contact with the substance to be cooled, or the closed cycle machine; and (2) those in which the cold air is exhausted from the machine and circulated through the cold stores in which the goods to be refrigerated are lying. These latter are known as the open cycle machines, and are the only ones now in use, the former class having become obsolete many years ago.

It is undoubtedly largely due to the cold-air machine that modern England owes her abundant and good supply of frozen meat, for, though the trade is now carried on largely by other systems of refrigeration, it was the cold-air machine which originated it and made it possible. In the cold-air machine ordinary atmospheric air is compressed to a pressure of about 50 lbs. on the square inch; the air by being compressed is raised in temperature, and the heat is taken out of it by passing it through a cooler, the tubes of which are surrounded by water. From the cooler the air is passed through a dryer, which consists of a series of tubes placed in, and surrounded by, the cold air returning from the cold chamber. In the dryer the compressed air is chilled and any moisture it contains is deposited. The air, still under compression, passes from the dryer to the expansion cylinder, where it expands (and in doing so helps to drive the machine); and, as air expanding while doing work is reduced in temperature, it leaves the cylinder at a temperature of a very considerable number of degrees below freezing-point, and goes away through air-trunks to the cold room. The dryer is a most important part of the apparatus. In the early days the principal difficulty with cold-air machines was the fact that when the air was expanded in the expansion cylinder, and reduced in temperature, the moisture contained in it was deposited in the form of ice in the cylinder and on the slide valves, choking them or raising them off their seats, and so rendering the cylinder inoperative. The invention and introduction of the dryer overcame this difficulty, and allowed the machine to work smoothly and well.[1]

Unfortunately the cold-air machine is not economical for large powers, though exceedingly useful for small powers and in small sizes. Professor Ewing[2] states that when examining the subject he was unable to find any case of a cold-air machine having a coefficient of performance greater than three-quarters of a unit, while the coefficient of a modern vapour compression machine is, under ordinary working conditions, between four and five units or about six times as great; which means that for every thermal unit put into the steam end of a cold-air machine only three-quarters of a unit are extracted from the cold room, while with a vapour compression machine from four to five units[3] can be extracted for the same expenditure.

The choice of a refrigerating agent is governed by its boiling-point at ordinary pressures; the latent heat of vaporization per pound; the number of cubic feet of vapour or gas that must be compressed to produce a certain refrigerating effect, or, in other words, the size of the compressor necessary; the pressure required to produce

[1] Dryers are not always used, as it is now found that by employing a very large water-cooling surface and by using the air over and over again, the deposition of moisture in expansion cylinder parts is not sufficient to interfere seriously with the working of the machine.

[2] *Howard Lectures*, Society of Arts, 1897.

[3] This, of course, depends on the temperature range at which the machine is working.

liquefaction of the gas at certain temperatures; and the specific heat of the liquid.

The qualities of the three chief refrigerating agents are set out in the following table[1]:—

| | Pressure in lbs. per square inch at 0° F. | Heat of vaporization per lb. at 0° F. | Volume in cubic feet per lb. at 0° F. | Specific heat of liquid. | Heat of vaporization per cubic foot. | Relative volume of compressor for equal refrigeration. | Loss due to cooling liquid. |
|---|---|---|---|---|---|---|---|
| | | | | | | | p.c. |
| Sulphurous acid, . | 10 | 171·2 | 7·35 | ·41 | 23·3 | 61·70 | 0·24 |
| Carbonic acid, . | 310 | 123 2 | 0·277 | 1·00 | 44·7 | 3·24 | 0·81 |
| Ammonia, . . | 30 | 555·5 | 9·10 | 1·02 | 61·7 | 23·3 | 0·18 |

All these substances have a boiling-point much below the normal. Ammonia, for instance, at atmospheric pressure becomes liquid at a temperature of $-30°$ F. Before it can be turned from a gas into a liquid, it has to be robbed of its latent heat, and it naturally follows that if its latent heat is restored to it, it will begin to boil and again become a gas. In boiling it will absorb heat from any substance that may be near it, and on being compressed the gas will, in its turn, deliver up the heat so absorbed to the cooling water trickling over the condenser, and again become a liquid.

The class of machine using a volatile liquid is separated into two divisions:—(*a*) the absorption machine, and (*b*) the compression machine. The absorption machine depends for its working on two substances that have an affinity for each other. When the machine is

[1] J. E. Siebel, *Compend of Mechanical Refrigeration.*

not at work these substances are united, but their nature is such that on the application of heat their state is changed, and they can be separated from each other. The principal substances used in absorption machines are either sulphuric acid and water or ammonia and water. It is well known that ammonia and water have a great liking for each other, and, if allowed to, will readily unite; but if heat from, say, a steam-coil, is applied to them when in their united condition, they will separate, and the ammonia will be driven off as a vapour under pressure. The vessel containing the ammonia and water and the steam-coil is called the generator, and as the ammonia vapour driven off is not entirely free from water, it has, in order to render it free from water or anhydrous, to be passed through an upright vessel called an analyser, containing a series of iron trays which catch any watery vapour that may have been carried off and return it to the generator, after condensation, as water. The ammonia vapour, now very nearly free from water, passes to the rectifier, where the remaining watery vapour is condensed and separated, and then to the condenser, where it is liquefied. From the condenser it passes through a regulating valve to the refrigerator, where the pressure is relieved and it re-evaporates. In doing so it absorbs heat from the brine, water, or other substance which circulates round it, and which is to be cooled. The ammonia gas which has evaporated is reabsorbed by the water from which it was driven off and is returned to the generator by a small pump. This small pump is the only moving part of an absorption machine, and requires very little power to work it. In some absorption machines even the pump is dispensed with and there are no moving parts.

The absorption machine is more economical than the cold-air machine, but not so economical as the compression machine. Professor Ewing has found that the absorption machine is from two and a half to three and a half times

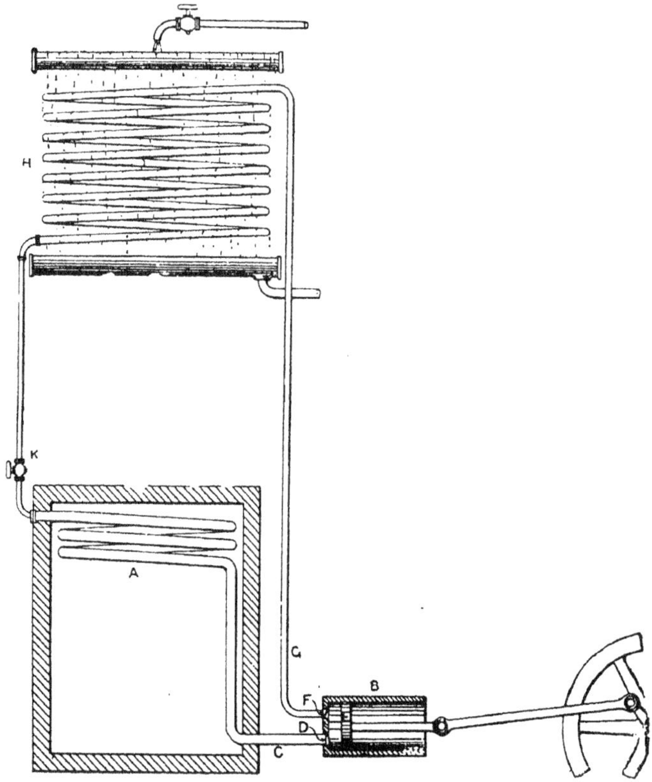


FIG. 1.

as efficient as the cold-air machine, and that the compression machine is from five to six times as efficient as the cold-air machine, and from two and a half to three times as efficient as the absorption machine. Fig. 1

shows diagrammatically the general arrangement of a compression refrigerating machine, using ammonia or some other volatile liquid. Assuming that the machine is working with ammonia, the gas from the refrigerating or expansion coils *A* is drawn into the compressor *B* through the pipe *C* and the valve *D*; on the return stroke of the compressor piston *E*, the valve *D* is closed and the gas is compressed and driven through the valve *F* into the pipe *G* leading to the condenser *H*. The ammonia gas in the suction pipe *C* is at a pressure of about 27 lbs. on the square inch, and at a temperature of 14° F. It contains the heat which has been absorbed from the atmosphere in the cold chamber, and which was necessary to turn the ammonia from a liquid to a gas. By being compressed this gas is decreased in bulk, and the heat contained in it is, so to speak, concentrated, so that the gas in the pipe *G* is at a temperature much above that of ordinary cold water. The condenser *H* has cold water trickling over it; this water is at a temperature of about 60° F., and absorbs the heat (representing the latent and specific heat of the ammonia and also the heat due to the energy of the engine) from the gas inside the coils and carries it away. The gas, robbed of its heat, can no longer remain a gas, and becomes liquid in the same way that steam, if cooled, will condense. Though a liquid, it is obvious that the ammonia will be ready to start boiling and again turn to a gas as soon as the pressure upon it is relieved. This is done by passing it through a minute opening in the expansion valve *K*. When the ammonia has passed through this valve it starts boiling and drops many degrees in temperature. Before it can boil, however, it must absorb heat, and as 1 lb. of ammonia requires, at atmospheric pressure, 555½ heat units before it can turn into a gas, these heat units have to be abstracted from the substance surrounding the expansion coils *A*, which, as a result, rapidly become covered with snow.

It will be seen that the cycle of operations consists of three parts:—

1. Compression or reduction in volume, and increase in temperature.
2. Condensing or giving up heat, and liquefying.
3. Expansion (boiling) or absorbing heat, and so gasifying.

In practice the pressure in the expansion coils is kept at from 1½ to 30 lbs. (corresponding to a temperature of from $-22°$ to $17°$) by the suction of the compressor, which is continually drawing off the gas as it is formed; but on account of the desirability of keeping $T_1$ and $T_2$ close together, the pressure, and so the temperature, is kept as high as possible consistent with the work the machine is doing. It is apparent that if the liquid ammonia is constantly fed into the expansion coils *A* through the valve *K*, and the gas drawn away by the compressors, a great amount of heat will be absorbed by the boiling ammonia in the coils, and that if these coils are placed in an insulated chamber, as shown in the diagram, that chamber will be rapidly reduced in temperature or refrigerated. The same remark, of course, applies to an ice-tank, in which the expansion coils are surrounded by brine, which is reduced in temperature, and so freezes the water contained in cans suspended in it.

The cycle above given applies to all of the vapour compression refrigerating machines; the details vary with different makers, and, of course, the size and strength of the various parts vary with the peculiarities of the particular refrigerating agent used. The relative volumes of the compressors are shown in the table previously given, and it is the size of the compressor necessary to produce a certain refrigerating effect that has so largely discounted sulphurous acid as a refrigerating agent. Thus it is that the ammonia and carbonic acid compression machines now practically hold the field.

# CHAPTER I.

## THE LAWS OF HEAT, FLUIDS, LIQUIDS, GASES, AND VAPOURS.

THE refrigerating machine is essentially a heat-engine reversed, and a proper knowledge of the physical laws that regulate the behaviour, under various conditions, of the gases and vapours which are employed in it is necessary before its action can be properly understood.

### HEAT.

All substances are capable of existing in one of three forms, and can be a solid, a liquid, or a gas, the change from one form to another being brought about either by the addition to, or withdrawal from, the substance of heat which is the source of all energy. All material substances, in whichever of these three forms they exist, consist of an infinite number of separate particles or molecules. Heat is a form of motion of these particles, and though the energy of this motion may be transformed and transferred, it can never be destroyed. When a body is in motion, it is said to have kinetic energy, and heat is represented by the kinetic energy or motion of the molecules of a body or substance; the greater the heat, the greater the motion, and *vice versâ*. When, however, a body, by the transference of its heat to some other body, has been reduced to the temperature of absolute zero (which will be referred to later on), heat is entirely absent, and the particles or molecules of the body are absolutely at rest and without internal energy of any sort.

**Temperature.**—The degree of heat of a body, or its thermal state, depends on the motion of its molecules, and is known as its temperature. This temperature is measured by the thermometer, the scales of which have been arbitrarily fixed—hot and cold being only relative terms—and depends upon the amount of energy the particles of the body possess.

**Transference of Heat.**—Heat can be conveyed from one body to another in three ways—either by radiation, conduction, or convection. The phenomena connected with the transference of heat are most complicated, and all these three processes may take place singly or at the same time.

**Radiation of Heat.**—Radiant heat is heat which can pass from one body to another through what is called the ether—that is to say, it is transmitted, like the heat of the sun, without the assistance of any material medium. It is always a transference of vibratory motion from a hot body to a cold, and the transference will continue until the source of heat is cut off or the temperatures are equalized. The passage of this radiant heat has no effect upon the temperature of the medium through which it passes, this remaining unaltered. The heat is also capable of being reflected or focussed, and in this respect is governed by exactly the same laws as light. It is, in fact, only when it becomes absorbed by the body upon which it falls that it is, strictly speaking, called heat; in the ether it is radiant energy.

**Conduction of Heat.**—Heat is conducted when it is

transferred from one body, or part of a body, to another by the means of some material medium or by actual contact. All substances have to a greater or less extent this power of allowing heat to flow through them. This property is called conductivity, and a substance that offers little obstruction to the flow of heat is called a good conductor, while a substance which offers great resistance is called a bad conductor or an insulator.

**Convection of Heat.**—The convection, or carrying of heat, takes place, as its name implies, when heat is bodily moved from one place to another in a fluid mass by means of the actual motion of that mass itself, such, for instance, as the carrying of heat by steam, heated air, or by hot water circulating in pipes. Heat can also be absorbed and reflected in the same way that light is absorbed by black or reflected by a bright surface. Generally speaking, light and heat follow the same law in this matter, and as a good reflector of light is a bad absorber, so a good reflector of heat is a bad absorber or radiator. An exception to this rule, which is of importance in cold storage work, is white lead, which, owing to its metallic nature, absorbs and reflects heat in an equal degree.

**Absolute Zero.**—At the temperature of absolute zero there is an entire absence of heat, and, as previously explained, the molecules of a body having lost all motion are at rest. The temperature at which this occurs is believed to be $-460°$ F. This figure has been determined because the expansion of a perfect gas under constant pressure is at freezing-point 1/492 of its volume for each degree change in its volume, and it follows, as a matter of course, that if it were cooled down to a temperature of 460° $(492-32)$ below zero Fahrenheit, its volume would become zero, and all motion would cease.

**Expansion by Heat.**—If molecular heat motion may be regarded as a form of vibratory motion of the particles of a substance, such motion of the individual particles would tend to separate them, and, obviously, if it were intensified, or, in other words, the temperature were raised, each particle would naturally require more space in which to move, and the whole warm mass would occupy a greater space than it did when cold. Accordingly, most bodies expand when heated and contract when cooled, the exceptions being water, and many metals near to but above their melting-point.

**Latent Heat.**—All substances can exist in one of three forms—a solid, a liquid, or a gas—and they may be changed from the one to the other by the agency of heat. When, however, this change is taking place, and a body is altering its molecular constitution, in passing from a solid to a liquid, or a liquid to a vapour, a certain amount of heat is required to increase the energy of the molecules, overcome the force of cohesion, and thus produce the change. In changing back from a liquid to a solid this heat is given up. In either case the heat does not appear sensibly—that is to say, it cannot be felt or measured by the thermometer—and it is, therefore, known as latent heat. At the same time this latent heat may be abstracted from the substances themselves, thus producing a fall in temperature, and this is what happens in the case of frigorific mixtures.

**Heat Capacity.**—The heat required to be absorbed by various substances before their temperature can be raised one degree varies with the substance, and the amount of heat thus required by any given substance is called its heat capacity.

**Specific Heat.**—The capacity for heat possessed by one

pound of water—*i e.*, the heat required to raise its temperature one degree—is always regarded as unity, and the capacity for heat possessed by any other substance in comparison to water is called its specific heat.

**Sources of Heat.**—There are several immediate sources of heat, the principal being Solar Radiation, Friction, Percussion, Mechanical Stress, Terrestrial Heat, Electric Action, and Chemical Combination (of which Combustion is a special case). It is apparent that these sources have originally derived their supply from solar radiation, as, for instance, in the formation of the coal measures.

**Chemical Combination.**—Heat formed by chemical combination may be best described as the heat that results when the constituent parts of one or more elementary bodies unite together chemically to form a body or substance entirely different from the original bodies. This definition is further explained by the fact that the newly formed substance possesses less energy than did the original component parts before they united. The difference of energy before and after the combination appears in the form of heat or electricity. When chemical compounds are again decomposed into their elementary parts, heat is absorbed, and the energy of the elementary parts is increased to its former intensity.

**Combustion.**—Combustion, the principal source of heat, is a chemical combination of carbon and oxygen (the carbon of the coal or coke with the oxygen of the air). This combination will only take place at certain temperatures, and before it can be started it is necessary to raise the temperature of the two bodies to the required point, which varies with different substances. The lower the temperature at which a body will begin to combine with the oxygen, and the more readily it will do so, the more inflammable it is said to be. The principal fuels are, of course, coal, coke, and lignite (new or partly formed coal). Coal contains from 55 to 80 per cent., coke from 85 to 97 per cent., and lignite from ·50 to 70 per cent. of carbon.

**Unit of Heat.**—Heat can be measured, and the unit of heat used in English-speaking countries is the British thermal unit (usually written B.T.U.). This is actually the amount of heat necessary to raise the temperature of one pound of pure water one degree Fahrenheit from a temperature of 39·1, which is the point of maximum density of water. For all practical purposes, however, it may be, and is, taken as being the amount of heat necessary to raise one pound of water one degree Fahrenheit, irrespective of temperature.

## Fluids.

A fluid is a substance which allows its molecules to be moved about to an indefinite degree by the exertion of the slightest force, and the difference between a fluid and a solid is that while a solid has rigidity, a fluid has none. Fluids are of two kinds—gases and liquids. This property of free movement of the particles in relation to each other is known as fluidity, and is possessed by gases to an even greater degree than by liquids. A fluid has the property of transmitting pressure exerted upon any part of it equally in all directions, and force transmitted through a fluid will act with equal pressure on all equal surfaces in a direction at right angles to those surfaces, as in a hydraulic press. It follows, therefore, that as pressure in a liquid is distributed in all directions, any substance floating upon water, and so creating a pressure upon it, will have a return pressure exercised upon itself, and will itself be borne up. This is known as buoyancy, and it is on account of this phenomenon that a substance weighed in water loses a portion of its weight equal to the weight of water it displaces.

**Specific Gravity.**—By this means the specific gravity of a body is ascertained by comparing its weight in air to its weight when immersed in water—that is to say, its specific gravity is the ratio of its density to that of water. For solids the standard is pure distilled water at 60° F., weighing 1000 oz. or 62½ lbs. per cubic foot, while for gases it is atmospheric air at the same temperature and 30 inches barometric pressure, weighing 31 grains per 100 cubic inches, or ·07 lb. per cubic foot. To ascertain the specific gravity of a solid, its weight in air is divided by the weight lost when weighed in water.

## Gases.

A permanent gas is an elastic fluid that cannot be liquefied, and it is exceedingly doubtful whether such a thing exists. Ordinary gases are those which will not liquefy at ordinary temperatures and pressures. All gases have weight—some, such as carbonic acid, being much heavier than air; and others, such as hydrogen, being much lighter.

**Compression and Expansion of Gases.**—If a gas is confined in a closed vessel and heat is applied, the energy of its molecules will be increased, the molecules will tend to separate, and the pressure will rise. If, however, heat is applied to gas in an unconfined state, it will expand freely and the pressure will remain constant, the energy which its molecules have derived from the heat being partly used in making room for the gas and giving it freedom to expand. The specific heat in the latter case is always greater than in the former.

**Resisted Expansion of Gases.**—When a compressed gas is

expanded while doing work, *i.e.*, behind a piston, a certain quantity of the heat it contained—the exact proportion depending upon the amount of work it is allowed to do—will disappear as work, and the temperature will fall. It is this property which makes the cold-air machine possible.

**Free Expansion of Gases.**—When, however, the pressure resisting the expansion of a gas is small, the gas is said to expand freely and does not perceptibly change its temperature. It has doubtless been noticed that when a refrigerating plant has been tested with air pressure, and the air is allowed to escape and expand from a pressure of, say, 300 lbs. per square inch to that of the atmosphere, there is no perceptible fall in its temperature; while if the same air was expanded while doing work in a cold-air machine or in a pneumatic drill or crane, its temperature would be many degrees below zero.

**Mixtures of Gases.**—Two or more gases, if brought together, will thoroughly mix if sufficient time be given them. The rate of mixing depends upon the weight of the respective gases, and is inversely as the square root of their densities—that is to say, a heavy gas like $CO_2$ will take longer to mix with air than a lighter gas. This rule is equally true and holds good for a mixture of any number of gases.

**Pressure of Gases.**—The pressure exerted upon the interior walls of a vessel or cylinder containing a mixture of gases under pressure is equal to the sum of the pressures which would be exerted if each of the gases occupied the same vessel alone. This is known as Dalton's law.

**Volume and Pressure of Gases.**—If a gas is heated while its volume remains constant, its pressure will increase proportionally to the absolute temperature; while if it is heated while its pressure remains constant, its volume increases according to the same law—that is, for a rise of temperature of 1° F. its volume increases by a constant fraction (1/492) of its volume at 32° F.

**Unit of Pressure.**—Pressure is measured in terms of the pressure of the atmosphere, which is taken as being equivalent to a pressure at sea-level of 14·7 lbs. per square inch. If this pressure of the atmosphere is removed by a vacuum pump and a vacuum formed, the vessel containing the vacuum is at once subjected to a pressure exerted upon it by the outside air in trying to flow in to destroy the vacuum. This property is made use of in suction pumps; the air in the suction pipe being removed by the action of the pump plunger, the water is forced up the suction pipe by the pressure of the air outside, and, as a pressure of 14·7 lbs. on the square inch will support a column of water 32 feet high, the water, theoretically, will rise to this level in the suction pipe. In actual practice, however, a height or suction head of from 20 to 25 feet is the maximum obtainable, as, owing to the more or less imperfect valves of the water-pump, a better vacuum can seldom be had.

**Liquefaction of Gases.**—All gases will become liquid, provided a sufficient pressure is applied and the temperature is sufficiently lowered. When gases or vapours are compressed, the energy or work of compression appears in the form of heat, and has to be removed before liquefaction will take place. In a compression refrigerating machine the heat removed by the condensing water is made up of the sum of the heat due to the energy of the steam-engine in compressing the gas and the heat that has been conveyed from the cold chamber or ice-tank by the refrigerating agent.

**Critical Temperature and Pressure.**—There is, however, a point called the critical temperature above which no amount of pressure will cause liquefaction. In the same way the pressure necessary to liquefy the gas at or near its critical temperature is known as the critical pressure. This latter term, however, is not so commonly heard as the former.

**Isothermal and Adiabatic Expansion and Compression.**—When a gas is expanded or compressed while its temperature remains constant, heat being added or taken away, it is said to be expanded or compressed isothermally; but when, owing to no heat being taken away or added, the expansion or compression is accompanied by a change of temperature—that is to say, the gas gets hotter or colder—the change is said to be adiabatic.

**Absorption of Gases.**—All liquids, some more than others, will, if left to themselves, take a certain amount of gas into chemical combination with themselves. During this absorption heat is given off, and if the gas has afterwards to be separated from the liquid, an equivalent amount of heat to that given off has to be added from an external source. This is the principle of the ammonia-absorption machine, the ammonia and water being allowed to combine, and then the ammonia being driven off by the application of heat.

## Vapours.

A vapour is merely a gas in a particular state, and is not a separate substance. When gases are lowered in temperature to their point of liquefaction, condensation begins to take place and they become charged with minute particles of liquid—in fact, become moist; in this state they are often called vapours. For instance, ordinary high and low pressure steam is a vapour, while superheated steam is a gas,

The term vapour is also applied to a gas below its critical point; thus, in this sense carbonic acid gas is a vapour at ordinary temperatures, while oxygen is not.

**Saturated Vapours.**—When the vapour is on the point of liquefying, it is said to be in a saturated state, and is then at its point of maximum density. Any vapour can, like steam, be superheated. It then becomes a gas, with all the peculiarities of a gas. All liquids will, if exposed to the atmosphere, absorb heat from it and evaporate, part of the liquid becoming a vapour. In a closed vessel the evaporation will continue until the pressure in the vessel, due to the formation of the vapour, is equal to the maximum density of the vapour for the particular temperature at which evaporation is taking place. This evaporation only takes place from the surface of the liquid.

**Boiling.**—When the temperature of a liquid becomes sufficiently high, vaporization will become rapid and take place from all points of the liquid, and what we know as boiling will occur. Bubbles of vapour will rise rapidly to the surface and there expand. The boiling-point of the various liquids varies, but boiling or ebullition will take place in any given liquid at a constant temperature for a constant pressure. If the vessel is a closed one, and the pressure is increased by the rapid formation of vapour, the temperature at which boiling will take place will rise, and as the pressure rises, more and more heat has to be supplied to the liquid before it will boil. If, on the other hand, the vessel is an open one, or so constructed that the vapour is drawn away or used as it is formed, as in a boiler supplying steam to a steam-engine, the temperature of the boiling-point will remain constant until boiling stops. If the pressure is removed or decreased, the temperature at which the liquid will boil falls correspondingly, the lowest boiling-points being obtained under the influence of vacuum. When no special pressure is stated, it is always understood that the pressure of the atmosphere is meant. This question of the relation of pressure and temperature is most important; it is the key to the economical working of all refrigerating machinery, and as such should be thoroughly understood.

**Latent Heat of Vaporization.**—When heat is applied to a liquid in order to turn it into a vapour, part of the heat, as we have seen, remains latent, and is expended in overcoming the cohesion of the molecules of the liquid. This heat is known as the latent heat of vaporization, and varies considerably with the pressure at which vaporization takes place. As the boiling-point of different liquids varies, that of some being much higher than that of others, it stands to reason that with a liquid having a boiling-point below the usual temperature of the atmosphere, the latent heat of vaporization, or the heat necessary to turn that liquid into a vapour, will be drawn from the objects surrounding it at the time of boiling. These objects, giving up their heat to the liquid, will fall in temperature and become cold or refrigerated, the amount of the refrigeration produced depending on the temperature at which the liquid is boiling. The vapour produced from the liquid contains not only the sensible heat, but also the latent heat of vaporization, and this heat has to be got rid of to some other object before the vapour can again become a liquid. If the boiling-point of the liquid is above the ordinary temperature of the atmosphere, this heat can be readily disposed of to the atmosphere or to water, and, as the boiling-point of liquid can be regulated at will within certain limits by the addition of pressure, all that is necessary, if the boiling-point is below the ordinary temperature, is, by the compression of the vapour, to raise its boiling-point to such a temperature that it will part with its latent heat to the atmosphere or to water.

**Permanent Gases.**—Permanent gases have been defined as gases that cannot be liquefied. Gases like oxygen, nitrogen, and hydrogen[1] were thought to come under this head, but, if subjected to sufficient pressure and cooled to a sufficient degree, permanent gases part with their latent heat and become liquid. As a matter of fact, all of the so-called permanent gases except helium have now been liquefied in this way.

[1] Professor Dewar succeeded in producing liquid hydrogen in 1898. It is by far the coldest liquid known, boiling at a temperature of $-454°$ F. It has also been frozen.

# CHAPTER II.

## THERMO-DYNAMICS AND ENTROPY.

**The Two Laws.**—It has been pointed out that heat is a form of energy which can be transformed or transferred into other kinds of energy, but which cannot be destroyed. The science which treats of the laws governing this transformation of heat into mechanical energy is thermo-dynamics, and is of great importance.

The first law of thermo-dynamics teaches us that heat and mechanical energy are mutually convertible. When work is turned into heat, or *vice versâ*, every British thermal unit requires for its production, and produces or is capable of producing by its disappearance or transformation, 778 foot-pounds of mechanical energy. This is known as Joules' equivalent.

Although heat and work, when convertible, are equivalent to one another, this conversion can only take place in all cases as far as the conversion of work into heat is concerned. It has doubtless been observed that though heat passes readily from bodies of a high temperature to those of a low, the reverse action never occurs, and it is impossible to convert the whole of any given quantity of heat into work, the reason being that heat cannot produce mechanical work without suffering a fall in temperature. During this fall only a portion of the heat is converted into work. The rest disappears as heat at a low temperature, and becomes unavailable for use, so that what is really used is not the energy contained in the heat, but its availability, and though energy may exist, it is of no use if non-available. This explains the second law of thermo-dynamics, which says that the unavailable heat of a cold body cannot be conveyed directly to the available heat of a hot one, unless under peculiar circumstances, with which the doing of work at some period or other is inseparable.

**Heat and Work.**—When a substance is heated its molecular activity is increased, and, though the conversion of heat into work can take place in many ways, the most common method is to take advantage of the pressure produced by the molecular activity of a gas due to its having been heated. If gas in this state is introduced behind a piston, the molecules will, in their efforts to get away from each other—that is to say, to each occupy a larger space—exert a pressure upon the piston and move it forward. At the same time as the piston moves and does work, the energy of the molecules will become less, and the temperature and pressure of the gas will fall. The doing of work is essential and, as has already been stated (Free Expansion), there will be no perceptible change in temperature unless work is done by the piston in moving.

A gas can be compressed or expanded in many ways, the most important being either without the subtraction or addition of heat, when its temperature will be changed (adiabatic), or with the loss or gain of heat, to such an amount that its temperature remains constant (isothermal). When heat is expended in doing work on a piston in a cylinder, it enters the cylinder at a high temperature and leaves it at a low one; the heat which is discharged as unavailable for use is less than the heat which entered, by the heat equivalent to the mechanical work done in moving the piston. The heat-engine, however, is theoretically a reversible machine, and the mechanical work done is the exact amount of work which, if the machine were reversed, would suffice to restore the heat discharged as unavailable to its original temperature, but no more; for, were it to raise the heat to a higher level of temperature than it was originally, the second law of thermo-dynamics would be violated.

**Adiabatic Expansion and Compression.**—If the heat in a reversible heat-engine was restored to its original condition without the addition of external heat, adiabatic expansion and compression would have taken place, the expansion being the adiabatic conversion of heat into work, the compression the adiabatic conversion of work into heat.

**Isothermal Expansion and Compression.**—If, on compression, the temperature is to remain constant, the heat expended in imparting energy to the piston must be replaced by heat drawn from some external source. This may also be reversed, and the work done by the piston be again converted into heat, by making it force the piston back and so compress the gas. Now, when work is done by the piston, the heat which was taken in from an outside source in order to keep the gas at a constant temperature will again appear as heat, and will, therefore, have to be removed to some outside source before the temperature of the gas will become what it was at first. If this cycle is carried out, and the gas is restored to its original pressure without change of temperature, isothermal expansion and compression will have taken place, the expansion being the isothermal conversion of heat into work, and the compression the isothermal conversion of work into heat, and, as everything has been restored to its original condition, no energy has been gained or lost.

**Unavailability.**—Though by this strictly reversible method we obtain the maximum conversion of heat into work, and *vice versâ*, it is obviously of no use for a continuous conversion, for if the work produced from heat has again to become heat, no energy is available for outside application, and we are no better off at the end than when we started. Heat cannot be converted continually into work unless the gas (or other working substance) undergoes a fall of temperature in the process. Even then the whole of the heat imparted to the medium is not available for use as work, however low the temperature to which it falls, unless absolute zero is reached, and as this is practically impossible, there is still molecular energy left in the medium when it is discharged. If, therefore, we heat the medium from some source of heat, such as coal, this unavailable energy has to be discharged to some substance of a lower temperature than itself, such as the atmosphere or water. It naturally follows that the contrivances or engines which make use of the heated medium will be working under the most perfect conditions when they are abstracting the greatest amount of heat from the medium during its fall from the high temperature to the temperature of discharge, and that if this process is reversed, and heat is raised from the low temperature to the high temperature, the greatest efficiency of the machine will be obtained when the heat is taken in at the highest possible low temperature and discharged at the lowest possible high temperature.

**Carnot's Cycle.**—In any cycle the medium always returns to the same condition as that from which it started, and so may be used continuously over and over again.

Carnot devised an ideal cycle which is reversible. This reversibility is the test of perfection in a heat-engine, but, to obtain it, every expansion of the medium must be resisted in such a way that by reversing operations the gas may be again compressed. In other words, there must be no free expansion, and heat must never be allowed to pass between bodies at sensibly different temperatures. In Carnot's ideal cycle, it is assumed that there was a cylinder containing a piston, and a given volume of gas as a medium, and that the cylinder and piston were incapable of receiving or giving up heat except at one particular place, where they could be brought into contact either with a source of heat or a cold body at will. Suppose them to be in contact in the first place with a source of heat, and then follow the cycle. First the piston is moved forward, and the gas expanding draws its heat from the source of heat, so that its temperature remains constant (isothermal expansion). Now the supply of heat is cut off, and the piston is moved forward again in the same direction. The gas expands while doing work, and falls in temperature (adiabatic expansion) until it is of the same temperature as the cold body to which it is now connected. The piston is now forced inwards in the reverse direction, and, as the gas is compressed, the heat of compression is withdrawn by the cold body, and the temperature remains the same (isothermal compression). The cold body is now cut off, the piston still further forced down, and the gas, being compressed, is raised in temperature until it again reaches the temperature of the source of heat (adiabatic compression), and the process is complete. This is the cycle for the heat-engine, but it can be readily reversed, and the gas allowed to expand at the low temperatures—in this stage taking heat from the cold body—and then being compressed adiabatically till its temperature reaches the higher value, and so on. This would be a cycle for the refrigerating machine, the two being as follows:—

| *Heat-Engine.* | *Refrigerating Engine.* |
| --- | --- |
| 1. Piston rises, gas absorbs heat from hot body and remains at constant temperature (isothermal expansion). | 1. Piston rises, no heat is absorbed and temperature falls (adiabatic expansion). |
| 2. Piston rises, no heat is absorbed, gas expands, temperature falls (adiabatic expansion). | 2. Piston rises, gas draws heat from cold body, temperature remains constant (isothermal expansion). |
| 3. Piston falls, gas gives up heat to the cold body, temperature remains the same (isothermal compression). | 3. Piston falls, no heat is absorbed and temperature rises (adiabatic compression). |
| 4. Piston falls, no heat is absorbed and temperature rises to original (adiabatic compression). | 4. Piston falls, gas gives up heat to the hot body and remains at constant temperature (isothermal expansion). |

In a heat-engine the relation between the heat $H$ supplied to the engine and the work $W$ got out of it is as follows:—

$$\frac{\text{Work}}{\text{Heat}} = \frac{\overset{T_1}{\text{The temp. at the stop valve}} - \overset{T_2}{\text{the temp. of the exhaust}}}{\underset{T_1}{\text{The temperature of the stop valve}}},$$

or supposing the pressure at the stop valve to be 100 lbs. on the square inch, and at the exhaust 15 lbs. on the square inch; the temperature of 100 lbs. steam in the former case is $337^\circ + 460^\circ = 797^\circ$ F. absolute. At atmospheric pressure it is $213^\circ + 460^\circ = 673^\circ$ F. absolute.

Therefore $\frac{797 - 673}{797} = \frac{124}{797}$, or as 124 is to 797. The consideration of this brings us back to the fact that heat cannot all be turned into work unless it is exhausted at the temperature of absolute zero.

The law reversed for a refrigerating machine would be :—

$$\frac{\text{Heat}}{\text{Work}} = \frac{\overset{T_0}{\text{The temperature of the refrigerator}}}{\underset{T}{\text{The temp. of the condenser}} - \underset{T_0}{\text{the temp. of the refrigerator}}}.$$

These formulæ show the fact that the refrigerating machine is the exact opposite to the heat-engine, and, as before stated, the greatest amount of refrigeration can be performed when there is the smallest difference between the temperature of the refrigerator $T_0$ and that of the condenser $T_1$. As this difference increases, the work becomes more and the refrigeration less.

**Entropy.**—The interpretation of heat transformations by means of a p.v. or indicator diagram is probably familiar to most people. It is only one of the methods that may be employed; but it has at least this advantage, that the diagram can be obtained automatically by means of the indicator, which enables the actual transformations taking place to be ascertained while the engine is working. The entropy diagram is another very important method of representation. At the end of a reversible cycle the entropy—that is to say, the quantity of heat taken into a system divided by the absolute temperature at which it enters—has the same value as at the beginning. This may be expressed otherwise by saying that if only the changes of entropy taking place along reversible paths are considered, its value at any one state must always be the same. It can be taken then as being one of the variable quantities defining the state of the body, and it can be shown that the state is completely defined if the temperature, entropy, and pressure thereat are known. Plotting absolute temperature ($T$) and entropy ($\phi$) against one another, the diagram (fig. 2) is exceedingly simple. During either of the isothermals, since the *temperature* is constant, each will be a straight horizontal line (*AB* or *DC*). During each of the adiabatics, since no heat enters or leaves, at any point of each the *entropy* will be constant; each is therefore a straight vertical line. And these statements are true *whatever may be the substance whose properties are*

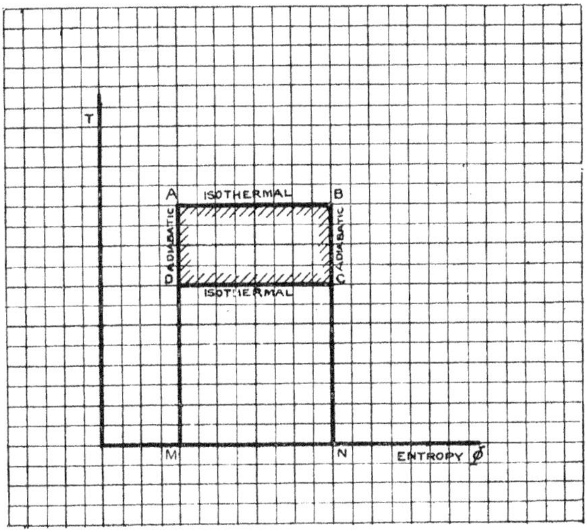


FIG. 2.

*so represented.* (The p.v. cycle would be different for every substance.) This in itself is an immense simplification. Again, the heat taken in during the upper isothermal will be change of entropy × absolute temperature—*i.e.*,

$$AB \times AM, \text{ or the area } ABNM.$$

Similarly the heat given out along the lower isothermal is represented by the area *DCNM*, and the external work, which is the difference of these heats, will consequently be the rectangle *ABCD*; and the efficiency is

$$\frac{Q_1 - Q_2}{Q_1} = \frac{AB(AM - DM)}{AB \times AM} = \frac{T_1 - T_2}{T_1}.$$

Nothing could be simpler, and if in practice reversible

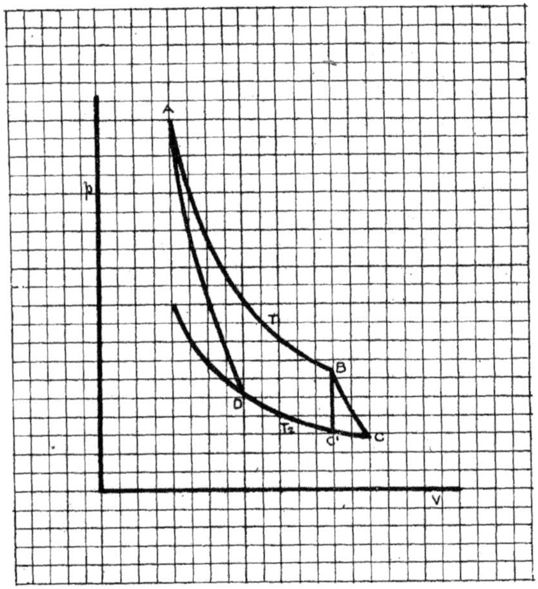


FIG. 3.

processes only had to be considered, any cycle would be represented by a rectangle on the diagram; and conversely, any such rectangle (with its sides parallel to the lines of reference) would represent a possible reversible process.

However, no process that is employed is strictly of this kind. A source of heat may be allowed to cool spontaneously without doing the work it might be made to do, and the peculiar fact is that if the chance of obtaining work is allowed to slip, it is gone for ever. Availability can only be restored by doing work upon the substance, and the availability thus restored is just equal to the least external work theoretically necessary to restore it, and therefore nothing is thereby gained. Since in any engine heat is

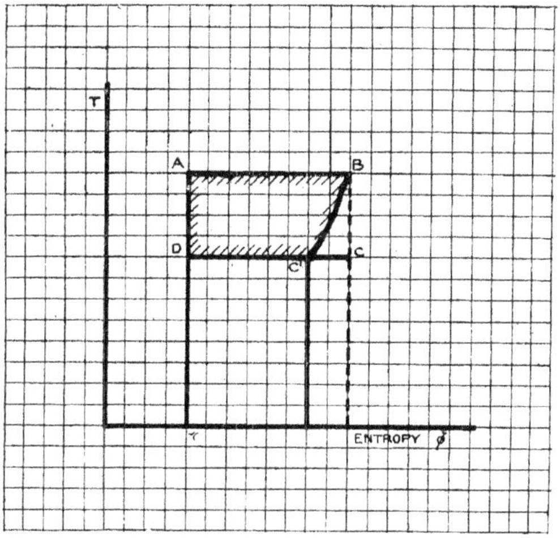


FIG. 4.

always being transferred by conduction and radiation between the parts of the machine which are at different temperatures, this source of loss of efficiency is always present. Consider the cycle in figs. 3 and 4, which is, in part, of this irreversible kind.

If, instead of being expanded adiabatically along $BC$, the substance be allowed to cool at constant volume along $BC^1$, it will do no work along this path. The work done in the whole cycle is represented by the area $ABC^1D$; this is less than the area $ABCD$, which represents the reversible case. In other words, an amount of heat might have been converted into work given by the area $BCC^1$; but instead of being converted, it has been transferred from the system as *heat* during the cooling from $B$ to $C^1$. The efficiency is therefore necessarily reduced.

**Internal Energy.**—It will be noticed that in the p.v. or indicator diagram the external work done *in any part* of a cycle is explicitly shown by means of an area, while in the entropy diagram it is the heat-change which is explicitly represented; in the former case the heat-change does not appear, and in the latter case the same statement is true with regard to the external work. It is this property of an entropy diagram which makes it so important in discussing refrigeration, because in such a discussion it is necessary to calculate the heat-change in any part of a cycle, and this cannot be done with an indicator diagram alone. The reason will be clear when it is pointed out that in the transformation the store of intrinsic energy which a body possesses may change during the transformation, owing to the performance of *internal* work. If the intrinsic energy is represented by $U$, then, in passing from one state to another, there must be

$$\text{Heat entering} = \text{Increase of } U + \text{External work done.}$$

It is only in passing round a complete cycle that $U$ recovers its original value, and, in consequence, the area representing balance of external work done (on the indicator diagram) represents also balance of heat entering, and that (in the entropy diagram) balance of heat entering represents balance of external work done. The intrinsic energy may change in several ways: if a change of temperature takes place, the *specific heat* is the increase of $U$ for one degree per unit mass; again, if vaporization takes place, it also increases, and what is called the *internal latent heat* is the increase in $U$ while unit mass evaporates. In any case, if the amount of heat necessary to effect a particular change is known, and we subtract from it the external work done during the change, the difference is the increase of intrinsic energy.

**Reversed Cycle and Principle of Refrigeration.**—If a reversible cycle is actually reversed, heat $Q_2$ will be taken into the working substance at the lower temperature $T_2$, and $Q_1$ given out from it at the higher temperature $T_1$; their old values will be retained, and since $Q_1$ is greater than $Q_2$, *work must be done upon the system* by an external agent equal in value to $Q_1 - Q_2$ in order to supply the deficiency. Such an arrangement may be used as a refrigerating machine, the amount of refrigeration being represented by $Q_2$, the heat which is taken from outside bodies at the lower of the two temperatures; while the heat $Q_1$ is rejected from the working substance to some other body (water in a condenser) at the higher temperature $T_1$. To cause the substance to pass through the requisite series of changes, it has been said that work must be done; this must, of course, be performed by an engine working in a direct cycle between a still higher temperature $T_1^1$ at which it takes in heat, and the intermediate temperature $T_2$. If it is supposed that both these machines are perfectly reversible, there is, if $W$ is the work performed by the direct engine upon the substance in the reversed one, and $Q_1^1$ is the heat it takes in at temperature (absolute) $T_1^1$,

$$W = Q_1^1 \frac{T_1^1 - T_1}{T_1^1}.$$

Similarly, considering the latter engine, there is external work done upon it $= W = Q_1 \frac{T_1 - T_2}{T_1} = \frac{Q_2}{T_2}(T_1 - T_2)$.

It will be well to examine before going further what is being effected here. A quantity of heat $Q_1{}^1$ is taken into the combined mechanism at one temperature $T_1{}^1$, and another quantity $Q_2$ is taken in at a second temperature $T_2$, while a quantity equal to the sum of these two (since

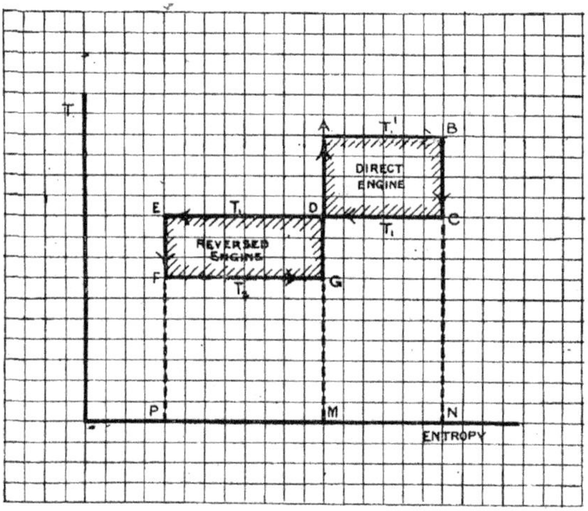


FIG. 5.

no external work is done by the combination on the whole) is rejected at an intermediate temperature $T_1$. What benefit has accrued from these transfers? Simply that a small quantity $Q_1{}^1$ so employed is able to lift a larger quantity $Q_2$ from the *coldest* of the three bodies to such a temperature that it may be moved out of the way by conduction.

It must be borne in mind that this is a result which, as has been shown, can only be brought about by aid of mechanical work. The relation between $Q_2$ and $Q_1^1$ can be found :—

$$W = Q_1^1 \frac{T_1^1 - T_1}{T_1^1} = \frac{Q_2}{T_2}(T_1 - T_2),$$

$$\text{whence} \quad \frac{Q_2}{Q_1^1} = \frac{T_1^1 - T_1}{T_1^1} \cdot \frac{T_2}{T_1 - T_2}.$$

If $T_1$ nearly equals $T_2$, or if $T_1^1 - T_1$ is great, $Q_2$ is large compared with $Q_1^1$; in other words, a small quantity of heat at a high temperature suitably employed is able to effect the transfer of a much larger quantity at a lower temperature from outside bodies. A diagram (fig. 5) will perhaps make this clearer: it shows the cycles performed by the direct and reversed engines; the areas of these cycles are to be equal, because they represent the work $W$ done by or upon the two engines.

$$Q_1^1 = \text{area } ABNM \quad . \quad Q_2 = \text{refrigerating effect} = FGMP.$$
$$= AM \times AB \qquad\qquad = FP \times FG.$$
$$\text{But } AD \times AB = EF \times FG,$$
$$\frac{Q_2}{Q_1^1} = \frac{FP \times FG}{AM \times AB}.$$

And this can be made large—the extreme temperatures remaining the same—by making $FG$ large compared with $AB$, *i.e.*, by making $AD$ large compared with $EF$. This then in brief is the principle of mechanical refrigeration. Heat can pass from hot to cold bodies spontaneously, but it will not pass from cold to hot bodies without some compensation; a direct engine will do the work necessary to take the heat and raise the body containing it to such a temperature that the heat will spontaneously flow away. The consideration of machines, however, will be restricted to the reversed section alone: this tacitly assumes that the best will be made of any means for *obtaining* external work. The performance of the refrigerator will be good if a large quantity of heat can be removed by a small amount of external work per cycle—that is to say,

$$\frac{Q_2}{W} = \frac{Q_2}{Q_1 - Q_2}$$

should be as large as possible: this ratio is called the coefficient of performance. Its value for any reversible cycle is $\frac{T_2}{T_1 - T_2}$.

Consideration of these formulæ will at once show that the smaller the range of temperature through which a substance is cooled, the better; and to cool a large quantity of air through a few degrees, say 10, will require a much smaller expenditure of energy than would be necessary if $\frac{1}{10}$ of the amount was cooled through 100°, though the actual amount of heat extracted is the same in each case. The moral is that if it is wished to cool a large quantity of air, it is theoretically[1] better to do it by the direct action of the refrigerating machine working through the 10° than to cool the smaller portion through the 100° and let it mix. This principle is clearly shown by the diagrams on p. 41, where three cycles are shown each of the same area, and with the lower temperature the same in each. Yet the refrigerating effects are very different, being equal to the dotted areas, and these are proportional to their horizontal widths, and therefore *inversely* proportional to the height $EF$ of the work-area, and this height is the difference of temperatures.

[1] With the cold-air machine this would not be possible in practice owing to the great size the necessary compressors would assume. It should be pointed out that this necessarily large range of temperature is one of the reasons why the cold-air machine is so inefficient.

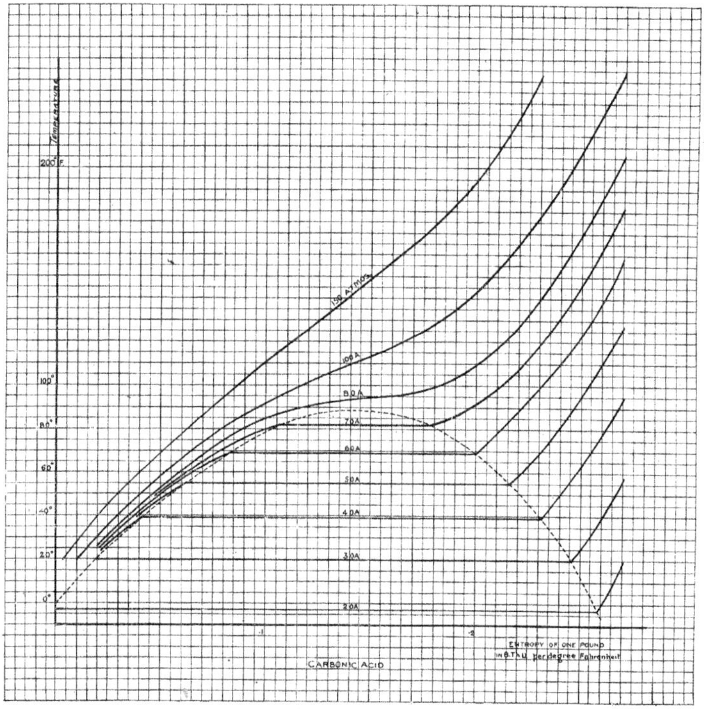


FIG. 11.

*To face page 40.*

The physical reason of this is that the external work is expended in restoring availability by raising the substance into which the heat pours along *FG* from the lower to the higher temperature. This restored availability is proportional to the increase of absolute temperature; and since a definite amount of work can only restore an equal amount of availability, the quantity of heat so raised

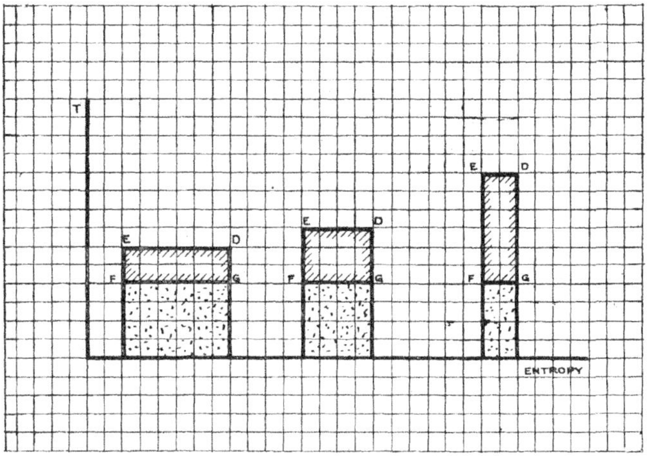


FIG. 6.

must vary inversely as the range through which it is raised.

Though the moral of this is clear, yet the result must not be pushed to extremes, for there obviously must be some difference of temperature left between the condenser (where the heat flows away) and the refrigerator (where it is taken in), otherwise the refrigeration becomes a fiction.

Machines in present use do not, however, follow this reversible cycle even approximately. Instead of the adiabatic expansion *AD*, which would require an expansion cylinder, the vapour employed is allowed to pass through an expansion valve. In doing so, it cools to the temperature of the refrigerator $T_2$. Now, although it does some external work in this stage, it does not do nearly as much as if the expansion cylinder were present. The amount it will do can be calculated in the following way.

Let a mass of gas be enclosed in a cylinder *A* at a pressure $p_1$, and the piston be moved gradually in such a way as to maintain the pressure constant while the gas issues through a nozzle into a second cylinder *B*, which also has a piston which is allowed to move so that the

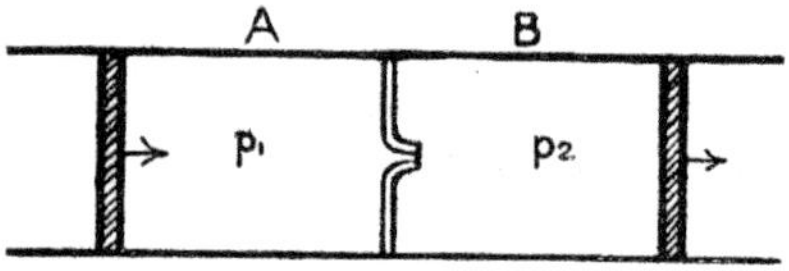


FIG. 7.

pressure in *B* remains at a constant value $p_2$; let the whole be so constructed that no heat can pass out of the double cylinder. If the state of the gas at two points distant from the nozzle is examined, the effect of the kinetic energy (which produces considerable modification near the nozzle where the changes of motion are great) may be neglected. Then, recalling the equation for the intrinsic energy, there is,

Heat entering $= O =$ increase of intrinsic energy $+$ external work done.

Now, while unit mass escapes through the nozzle if the motion is steady, the work done by the gas upon the second piston is $p_2 v_2$, where $v_2$ is the volume of unit mass at the pressure $p_2$ and temperature $T_2$: the work done by the first piston upon the gas is $p_1v_1$, where $v_1$ is the volume of unit mass on the first side. Hence

$$O = U_2 - U_1 + p_2v_2 - p_1v_1,^1$$

or $$U + pv \text{ is a constant.}$$

Since the pressure outside the jet is much less than that inside, the expanding gas does not do its full amount of work, and it is important to examine what the result of this is; it can easily be shown by the diagram (fig. 8) on p. 44.

Suppose that it transforms along the path $ARD^1$. (The form of this path is very various in different cases; everything depends upon the nozzle employed. Close to the nozzle the temperature is much lower than at a distance—*i.e.*, lower than $T_2$. Its actual shape, however, is not a matter of concern.) Now let $AD$ and $DD^1$ be an adiabatic and isothermal by which the same point $D^1$ might be reached—these represent a reversible path, therefore, between $A$ and $D^1$—the work done along them will equal the area $ADD^1N^1M$, while an amount of heat $q$ is taken in along $DD^1$. Now in the expansion through the expansion valve the work done is $p_2v_2 - p_1v_1$; that is,

$$\text{area } OQD^1N^1 - OPAM.$$

Therefore the work that might be done, but is omitted to be done, is the difference of these, *viz.*:—

$$ADD^1N^1M - OQD^1N^1 + OPAM.$$

[1] If a region close to the nozzle were examined where the gas is moving rapidly (velocity $= V$), it would be necessary to add the corresponding kinetic energy to the right-hand side. The equation would become $O = U_2 - U_1 + p_2v_2 - p_1v_1 + \frac{1}{2}V^2$. Since part of the total energy is represented by this kinetic form, there is less for thermal effects; the changes of temperature in the "rapids" may, in consequence, be very large compared with the changes in passing from one region to another, in each of which the velocity is small.

Now, by inspection of the diagram the first and last added together = area $OPADD^1N^1O$, and subtracting the area $OQD^1N^1$, there is left

$$\text{area } QPADD^1Q.$$

This represents the work which the expanding gas

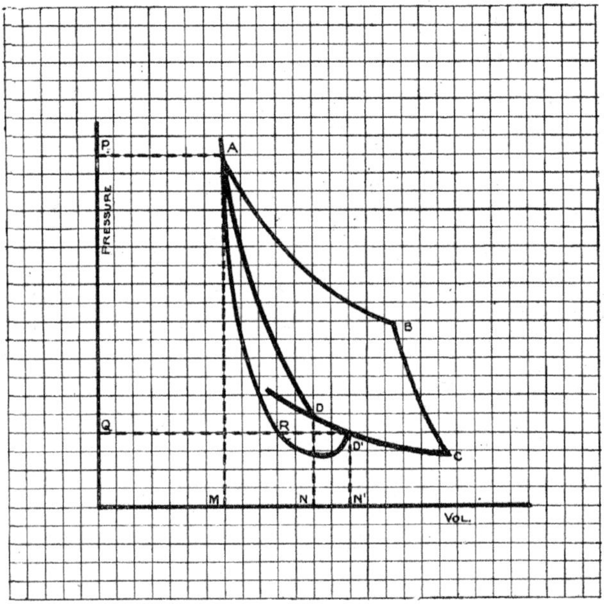


FIG. 8.

does *not* do; consequently the motor engine has to do it instead: this increases the quantity $W$ in the equation for the coefficient of performance. This, however, is not its only effect, for $DC$ is the isothermal along which heat is being extracted from the cooled bodies, and the portion of this isothermal which remains to be done is

$D^1C$ instead of $DC$; and less heat will be absorbed than if the gas had been compelled to expand adiabatically and then allowed to extract heat at constant temperature. This may be more completely understood if it is described in an alternative way. In the reversible path a quantity of heat $q$ is introduced into the system along the portion of the isothermal $DD^1$; the working substance can reach the same state $D^1$ *without the introduction of any heat* if a correspondingly less amount of external work is done along the new path, and this is exactly what takes place in the rush through the expansion valve. Thus the refrigerating effect per cycle—*viz.*, $Q_2$—is *diminished*, and this again diminishes the coefficient of performance. Thus the use of an expansion valve in place of an expansion cylinder has a two-fold deleterious effect; nevertheless, in consequence of the increased simplicity of apparatus, it is universally employed.

This may be further illustrated by means of a $T\phi$ diagram. The same letters in the two diagrams represent the same states; the areas now represent heats. $DD^1N^1N$, fig. 9, represents the *decrement* of refrigeration, and consequently the *additional* work to be performed by the motor; this latter is also equal to the area of the figure $ARD^1DA$.

The residual amount of possible refrigeration is $D^1CSN^1$, the total external work to be done is $ABCD^1RA$, and the coefficient of performance is the ratio of these areas.

Other divergences from the perfect cycle will be dealt with where they arise. The previous considerations will apply *whatever the working substance used*—whether gas, liquid, or solid. Practically the only substances used to-day are vapours which can be liquefied by pressure, such as ammonia ($NH_3$) or carbonic acid gas ($CO_2$). It will be necessary to examine some of the properties of these vapours, taking carbonic acid ($CO_2$) as an example.

**Properties of Carbonic Acid ($CO_2$).**—If the gas is operated on at 120° F., the relations between *p*, *v*, and *T* are not much unlike those for an ordinary gas. If the temperature is lower than this, greater and greater deviations arise.

If, for example, fig. 10, the gas is operated on at 70° F.,

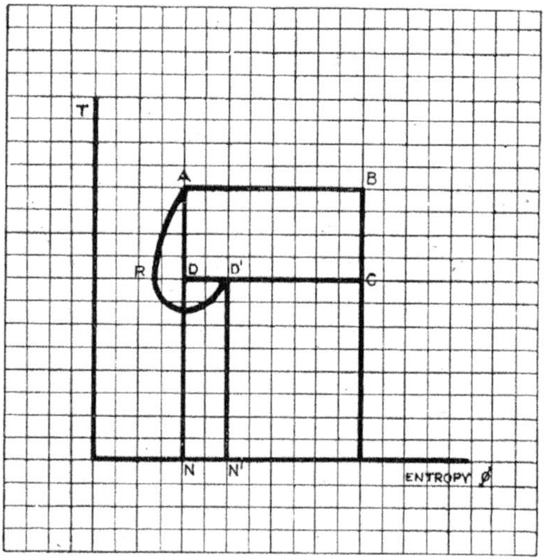


FIG. 9.

then, starting with a pressure of about 50 atmospheres and decreasing it, the volume changes so as to give rise to the curve *BA*; but if the volume is decreased (still keeping temperature constant), the gas *liquefies* and the pressure *keeps constant* until it has all become liquid. The representative curve is *BC*; when it is all liquid very great increase of pressure produces very little alteration of volume, and this portion of the curve *CD* is therefore very nearly a vertical straight line. If the gas is taken at a higher temperature (*e.g.*, 80° F.), the straight portion $BC$ becomes shorter, and this shortening goes on with rise of temperature until $t = 88°$ F., when it vanishes. The significance of this is that just below this temperature (88° F.) the gas changes to liquid without

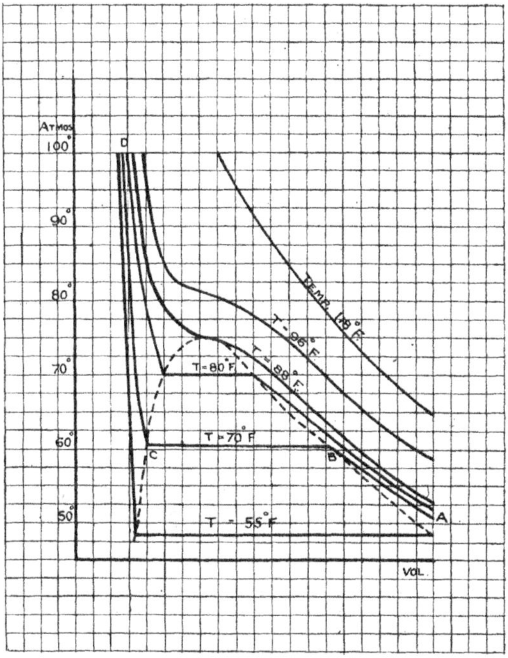


FIG. 10.

any sensible change of volume, while above this temperature, although the change of volume may be as great as that which would ensue from liquefaction, no liquefaction occurs. This temperature is called the critical temperature; the corresponding pressure and volume of unit mass are called the critical pressure and volume respectively. Their values for some substances are given in the following table:—

CRITICAL DATA.

| Substance. | Temp. F. | Pressure (atmospheres). | Volume in cub. ft. of 1 lb. |
|---|---|---|---|
| Carbonic acid gas, | 88 | 75 | ·035 |
| Water, . . | 690 | 195 | ·037 |
| Ammonia, . . | 266 | 115 | ·048 |

**Calculation of Entropy.**—If 1 lb. of the liquid rises in temperature, each degree rise corresponds to the reception of an amount of heat equal to the specific heat. While this evaporates an amount of heat is taken in equal to the latent heat (this is partly spent in doing external work owing to the change of volume which occurs). While the vapour rises in temperature after evaporation is complete again, the amount given per degree by the specific heat is found; but the value of this will depend upon the kind of change that takes place, the amount required to raise the temperature 1° being greater if the pressure remains constant than if the volume remains constant, owing to the external work done in the former case.

All these increments of heat increase the entropy of the substance: in the case of the latent heat the calculation of the consequent increase of entropy is easy because the temperature remains constant. If $r$ is the latent heat, its value will be $\frac{r}{T}$, where $T$ is the absolute temperature of evaporation. If $r$ is expressed in thermal units, and $T$ in degrees, $\frac{r}{T}$ is a mere number, and has the same value whatever system of units is adopted. The calculation of the increase of entropy during a rise of temperature is not so easy, because the temperature at which the heat is received changes as each fresh quantity of heat is added. If $\tau$ is a *small* increase of temperature, the quantity of heat producing it is $C\tau$, where $C$ is the specific heat. The corresponding increase in entropy is $\frac{C\tau}{T}$, where $T$ is the temperature at which it is received. The total increase while the temperature increases from $T_2$ to $T_1$ will be

$$\Sigma\frac{C\tau}{T},$$

where the whole rise is divided into small portions each equal to $\tau$. The value of $\frac{C\tau}{T}$ is found for each, and all are added together. In any case the calculation demands a previous complete study of the properties of the substance, and even then considerable labour.[1]

It has been done by Dr Mollier[2] for carbonic acid gas, and the complete chart (in which the units are changed throughout to British units) is shown in fig. 11. Calculations for ammonia, so far as the imperfections of the data known with regard to ammonia have per-

[1] It can be shown that the rate of increase of entropy of a vapour with pressure when the temperature keeps constant is equal to the rate of *decrease* of volume with temperature when the pressure keeps constant; a complete knowledge of the latter (which is merely the coefficient of expansion of the vapour) enables the former to be calculated; and this is the easiest mode of obtaining it.

[2] R. Mollier, "Ueber die Kalorischen Eigenschaften der Kohlensäure aussenhalb des Sättigungsgebietes," *Zeitschrift für die gesammte Kälte-Industrie*, vol. iii., 1896.

mitted, have also been made, and the corresponding entropy chart is shown in fig. 12. Great use will be made of these diagrams in the following discussion.

**Complete Discussion of Actual Refrigerating Cycles.**—The machine may be diagrammatically represented as in fig. 13. By means of a compressor the refrigerant (usually ammonia or carbonic acid) is forced round and round a

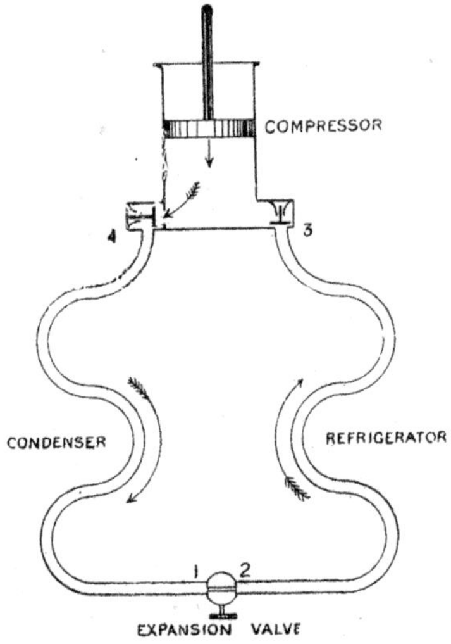


Fig. 13.

circuit of pipes. At a certain point an expansion valve is inserted, the consequence of which is that there will be a considerable drop of pressure as the refrigerant goes through it. Neglecting the fall of pressure as the refrigerant flows along the pipes themselves, the whole may be regarded as being divided into two parts, the refrigerant in the left half (the condenser) being all at one pressure $p_1$, while that in the right (the refrigerator) is all at

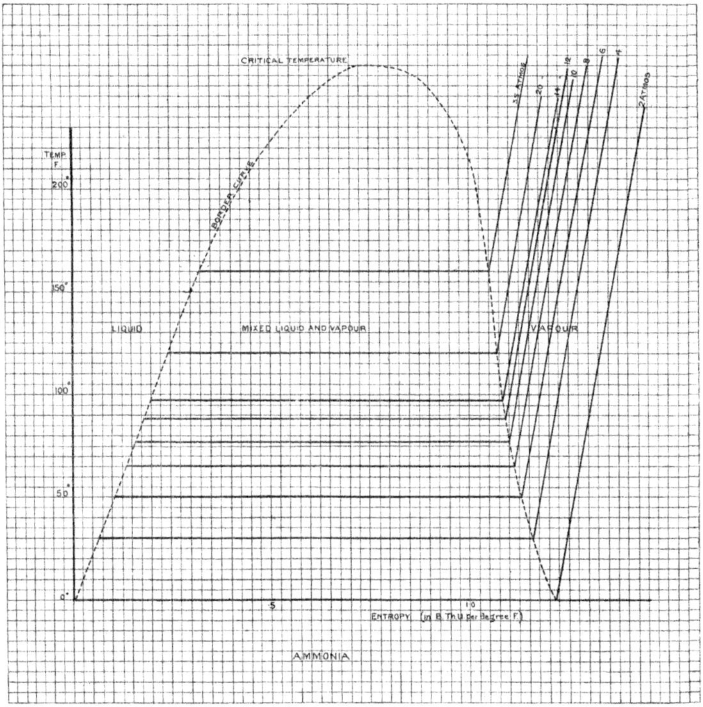


FIG. 12.

*To face page 50.*

another pressure $p_2$, this difference of pressure being maintained by the continued action of the compressor. The condenser is kept cool with water, and by the time the refrigerant reaches the position **1**, it may be supposed that it will have come to the same temperature. Its state will there be defined by the three quantities $p_1$ (pressure in atmospheres), $T_1$ (*absolute* temperature on the Fahrenheit scale), and $v_1$ (the volume in cubic foot of 1 lb. weight): it is always in the liquid form at this stage. In passing through the expansion valve the temperature always falls, and part of the liquid usually evaporates; hence in the position **2** the state is given by $p_2$, $T_2$, and $v_2$, where $p_2$ and $T_2$ are each less than $p_1$ and $T_1$, while $v_2$ is greater than $v_1$; $v_2$ is, of course, the volume of 1 lb. of *mixed* liquid and vapour in the proportion in which they exist at this stage. (As has been explained, this position must not be taken so near the throttle as to be affected by the rapids which exist there; neither should it be taken so far away from it that heat may have had opportunity of being received through the refrigerator walls; for it is supposed that the change from **1** to **2** is taking place without any access of heat.) While flowing through the refrigerator it is receiving heat, and more liquid evaporates in proportion; the state at position **3** will be denoted by $p_3$, $T_3$, $v_3$, and it is supposed to be still sensibly in this state when sucked into the compressor. The compression which ensues is supposed to take place adiabatically; the result is that the temperature and pressure are much higher at **4** than at **3**, while the specific volume is less: these will be denoted by $T_4$, $p_4$, $v_4$.

The changes of entropy that correspond can be obtained from the general chart if the values of the temperatures $T_1$ and $T_2$ and the pressures $p_1$ and $p_2$ are given. Considerable modification can be made in practice by adjustment of the expansion valve, and hence several cases arise. The chief variation is in the amount of liquid present in

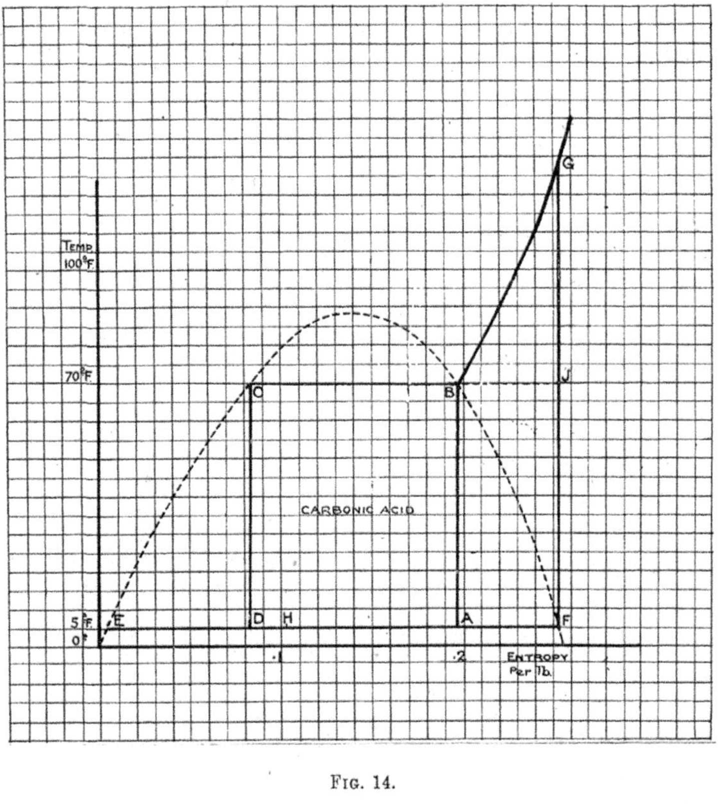


FIG. 14.

the vapour in position **3.** If it is all evaporated here it will be superheated at the end of the adiabatic compression (dry compression); any amount of liquid may be allowed to remain (corresponding to the priming in a steam-engine) by adjusting the expansion valve, and these give corresponding degrees of "wetness" in compression (wet compression).

All the inferences are obtained by measuring the diagrams. Carbonic Acid Machines will be first considered.

**Carbonic Acid Machines working without Superheating and with Expansion Cylinder.**—The case taken first is one in which the temperatures at **1** and **2** are 70° F. and 5° F. respectively. Their corresponding absolute values will be $T_1=530$ and $T_2=465$. The saturation pressures at these temperatures are 60 and 24 atmospheres approximately. If the vapour *never* superheats and is all just condensed into liquid when it reaches the expansion valve, the reversible cycle (which is taken as the standard) is the cycle *ABCD*, fig. 14. The fraction of liquid unevaporated at *A* (position **3**) is $\frac{AF}{EF}$ lb. (since the change of entropy is proportional to the mass evaporated, and the whole mass whose circulation is being followed is 1 lb.)—*i.e.*, about ·23 lb.—and after the adiabatic expansion *CD*, the amount $\frac{ED}{EF}$ lb. = ·31 lb. is already in the state of vapour. Consequently the amount left for producing useful refrigeration is (1 − ·23 − ·31) lb. = ·46 lb. The heat it removes is change of entropy × absolute temperature = area lying between *DA* and the *base line drawn through the absolute zero of temperature—i.e.*, . . . . . 53 B.T.U.
Work done by compressor = area *ABCD* = 7·41 „

$$\text{The coefficient of performance} = \frac{53}{7\cdot41} = 7\cdot15,$$

$$\left(\text{and this is ratio } \frac{465}{65} = \frac{T_2}{T_1 - T_2} \text{ as it should be}\right).$$

**Carbonic Acid Machines working without Superheating and with Expansion Valve.**—If an expansion valve is employed, the expansion will terminate at some point *H* instead of *D*, and its position may be determined by means of a p.v. diagram. [It is in practice taken such that $DH \times T_2 =$ area *EDC* on the entropy diagram; it can be shown that this always gives too low a value, and in many cases much too low. In this particular example the area $EDC = 2{\cdot}62$ thermal units, while the heat represented by *DH* (calculated by the exact rule) is 4·60 thermal units[1]—*i.e.*, roughly, twice as great.] Instead of 53·0 units available for refrigeration, there are now only $53{\cdot}0 - 4{\cdot}60 = 48{\cdot}40$ units. The work to be done by the compressor is increased by an equal amount—*i.e.*, it equals

$$7{\cdot}41 + 4{\cdot}60 = 12{\cdot}01 \text{ thermal units,}$$

and the coefficient of performance is reduced to

$$\frac{48{\cdot}40}{12{\cdot}01} = 4{\cdot}03,$$

[1] In the case of $CO_2$ the following approximate rule for calculating this is *very nearly* exact:—Add $\frac{1}{20}$ difference of pressure in atmospheres to the area *ECD* expressed in thermal units; the sum is the quantity of heat represented by *DH*. If the condenser temperature is above 70° F., this rule gives rather too low a value. In measuring quantities of heat by means of areas on an entropy diagram, it must be borne in mind that the area in square centimetres (obtained by adding the number of squares on millimetre paper or by a planimeter) must be multiplied by the *physical equivalent of* 1 *sq. cm.*; thus, if a horizontal centimetre represents ·02 unit of entropy, and one centimetre height represents 10° F., the *physical equivalent* of 1 sq. cm. is $\cdot 02 \times 10 = \cdot 2$ British thermal unit. The geometrical area must be multiplied by ·2 in order to yield the result in B.T.U. Again, in a p.v. diagram it must be remembered that if *p* is in lbs. per square foot and *v* in cubic feet, the areas, when reduced as above, express work in foot-lbs; to obtain it in thermal units one must further divide by 778 the mechanical equivalent of heat.

which is less than two-thirds the maximum possible. Each lb. of material circulating also removes less heat, which is an additional defect as we have previously indicated.

**Carbonic Acid Machines working with Superheating.**—Now take a case in which the liquid is just all evaporated in position 3. The corresponding position on the entropy diagram is $F$; it is then adiabatically compressed till the pressure is $p_1$—*i.e.*, it will pass along the vertical $FG$ till it reaches the continuation backwards of the constant pressure line $CB$ at $G$. Its temperature is seen (by diagram) to rise to 130° F., and at this temperature it enters the condenser, which is at 70° F. It consequently at once begins to part with its heat to the walls of the condenser, and thus a second irreversible stage is introduced—*i.e.*, a stage involving unnecessary performance of work.

(In fact, if the adiabatic compression had ceased at $J$ and then the vapour were compressed isothermally to $B$ and thence to $C$, the reversibility would be restored and less work performed by the amount $BGJ$; the last portion of the compression $JG$ is *wasted*, producing a deleterious rise of temperature, and the heat tumbles tumultuously out without doing its full share of work, when it gets into the condenser which is at a much lower temperature. This may be otherwise expressed by saying that we restore availability by doing external work upon the substance along $JG$, and then we wilfully let it fritter itself away while falling in temperature.) By allowing the whole liquid to evaporate in the refrigerator, the additional refrigerating effect, $AF \times$ abs. temp. $= 26{\cdot}04$, is gained in each cycle, but the compressor has to do *greater additional* work, *viz.*, $ABGF = 5{\cdot}20$. With superheating then to this amount, it is found that for the coefficient of performance

($\alpha$) with adiabatic expansion along $CD$,

$$\frac{53{\cdot}0+26{\cdot}04}{7{\cdot}41+5{\cdot}20}=6{\cdot}3,$$

while the refrigerating effect per cycle = 79·05 B.T.U. per lb. circulating.

($\beta$) With the expansion valve

$$\frac{48{\cdot}40+26{\cdot}04}{12{\cdot}01+5{\cdot}20}=4{\cdot}3,$$

while the refrigerating effect per cycle = 74·45 units per lb. circulating.

Reviewing these various cases, it will be seen that the performance is diminished by allowing superheating if the expansion cylinder is employed, whereas it is slightly improved by it if the expansion valve is employed, while the superheating increases the refrigerating effect per lb. in both cases.

**Ammonia Machine working without Superheating and with Expansion Cylinder.**—Taking next the case of an ammonia machine working at the same temperatures 530 and 465 abs. Fahr., the saturation pressures at these temperatures are 8·8 and 2·3 atmospheres.

If the vapour nowhere superheats and is just completely condensed into liquid when it reaches the throttle, the standard reversible cycle is the rectangle $ABCD$ (fig. 15). The weight of liquid unevaporated at $A$ (*i.e.*, the priming) is $\frac{AF}{EF}$ lb. = ·075 lb.; and after the adiabatic expansion $CD$, the amount $\frac{ED}{EF}$ lb. = ·081 lb. is already in the state of vapour, and therefore does not produce useful effect.

Consequently the amount left for producing useful refrigeration is

$$(1 - {\cdot}075 - {\cdot}081) \text{ lb.},$$

that is, ·844 lb. The heat it removes is equal to the consequent change of entropy × abs. temp. = ·97 × 465 = 451 B.T.U.

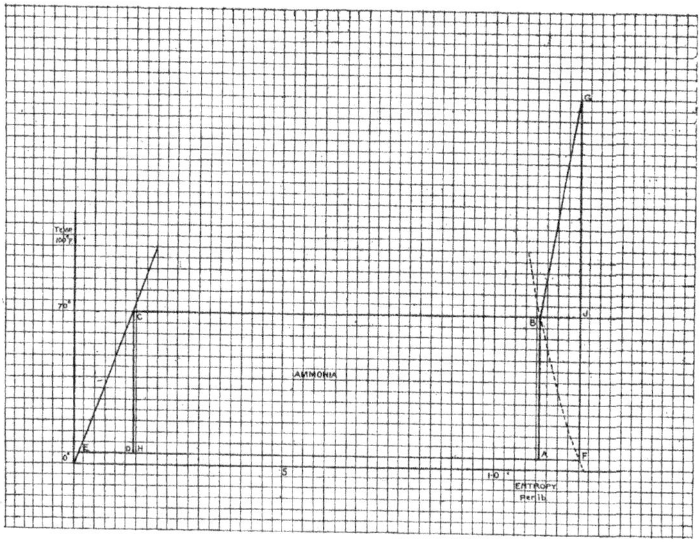


FIG. 15.

The work done by the compressor during the cycle is the area $ABCD = 63$ B.T.U., and the coefficient of performance is

$$\frac{451}{451-63} = \frac{451}{388} = 7{\cdot}1 \left( = \frac{465}{530-465} \text{ as it should be} \right).$$

**Ammonia Machines working without Superheating but with Expansion Valve.**—If an expansion valve is employed, it is necessary to determine the point $H$ representing the state in position **2**. Finding this by aid of a p.v. diagram, we obtain $DH = \frac{4 \cdot 85}{465}$ units of entropy,[1] which represent 4·85 B.T.U. of heat.

Consequently the useful refrigerating effect is diminished by this amount, and the work left for the compressor to do is increased by an equal amount. The coefficient of performance is reduced to

$$\frac{451 - 4 \cdot 85}{63 + 4 \cdot 85} = \frac{446 \cdot 1}{67 \cdot 85} = 6 \cdot 58.$$

**Ammonia Machines working with Superheating and Expansion Cylinder.**—If the vapour leaves the refrigerator in the just saturated state, it will become superheated in the compressor; at the end of compression its temperature will be about 170° F. The cycle is now $FGBCDF$ if adiabatic expansion were to take place along $CD$.

There is now additional refrigerating effect $AF \times 465 = 35 \cdot 0$ B.T.U. due to the complete vaporization that occurs, but an enlargement of the work-area equal to $ABGF$. This additional work is 7·2 B.T.U., and the coefficient of performance becomes

$$\frac{451 + 35}{63 + 7 \cdot 2} = 6 \cdot 9.$$

If the expansion valve is employed instead of the ex-

[1] The position of $H$ can be found with almost perfect exactness by means of the approximate formula $DH \times T_2 =$ area $ECD$ (in B.T.U.) $+ (\cdot 071 \times$ difference of pressure in atmospheres).

pansion cylinder, the coefficient of performance is evidently

$$\frac{446{\cdot}1+35}{67{\cdot}85+7{\cdot}2}=6{\cdot}4.$$

Thus, in the case of ammonia it does not make any great difference which of these systems be adopted: the disadvantage of the slightly reduced coefficient when superheating is permitted may be considered to be partly counterbalanced by the increased cooling effect of 1 lb. of circulating material. The cooling effect is produced at a rather greater expenditure of power, but it takes place faster if two machines circulating at the same rate are compared.

**Influence of Condenser Temperature.**—In the case of a $CO_2$ machine, in consequence of the lowness of its critical point (88° F.) great variations are caused if the condenser temperature be changed.

As an example, take the case in which $p_1=80$ atmospheres, $p_2=30$ atmospheres, and the refrigerator temperature is 20° F., fig. 16.

The cycles which result from various condenser temperatures are shown in the figure; the amount of vaporization which takes place in the expansion valve has been worked out for the several cases, with the following results:—

CONDENSER TEMPERATURES.

| | 40° F. | 60° F. | 80° F. | 100° F. | 120° F. |
|---|---|---|---|---|---|
| $DH$ | ·0054 | ·0070 | ·010 | ·025 | ·030 |
| $EH$ | ·02 | ·047 | ·080 | ·188 | ·224 |
| $x=\frac{EH}{EF}$ | ·098 | ·209 | ·355 | ·834 | ·994 |

The bottom row shows the fraction of a lb. which is rendered useless for refrigerating purposes due to being in the state of vapour on entering the refrigerator.

It will be seen that whereas if the condenser temperature is 40°, the amount per lb. thus lost is only about $\frac{1}{10}$ lb.;

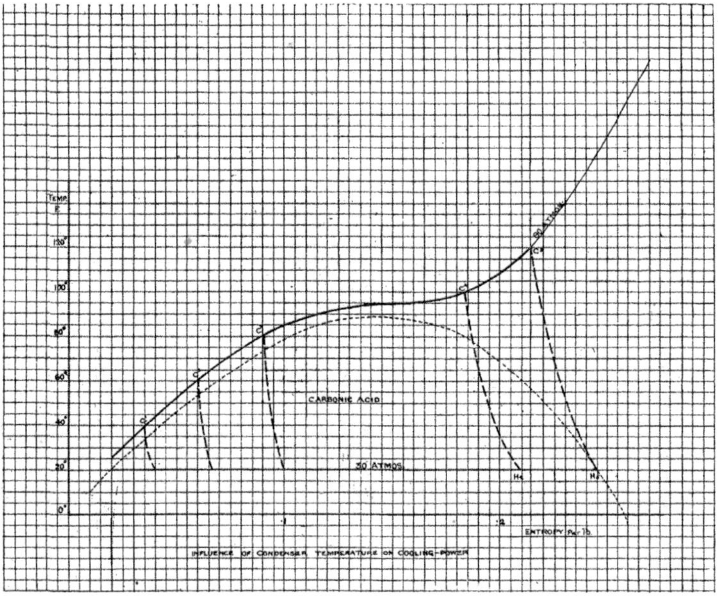


FIG. 16.

if the temperature is 120° F., practically the whole is lost and the refrigerating effect is *nil*. The loss increases rapidly when the temperature is a little over the critical temperature; this results (as will be very evident on inspection of the diagram) from the very flat form of the constant pressure curve in that region. It should be carefully noticed that the work-area diminishes at the same time, and hence the coefficient of performance remains large even past 80° F.; the deleterious action of the increased condenser temperature is that more material has to be circulated in order to remove the same quantity of heat.

In this example the vapour does not liquefy in the condenser at all, the pressure chosen being above the critical pressure. It might at first be thought that this would prevent any effective refrigeration from taking place. In the passage through the expansion valve, as the pressure and temperature decrease owing to the expansion, liquefaction does occur, and the latent heat of its future evaporation in the refrigerator produces, as usual, the effective cooling.

It must be understood that the above diagram and statements only apply to a special case, *viz.*, that in which the pressures are 80 and 30 atmospheres respectively; hence it must not be thought that it is intended to imply that 120° F. is the *ultimate limit* of workable temperature. Any case may be worked out by the help of the entropy chart; and inspection of this easily shows that if the higher pressure is increased there is a corresponding increase in the possible condenser temperature.

If superheating were not allowed it would be impossible to work with the condenser near the critical temperature. This will be obvious from fig. 17, representing two cycles in which the vapour is throughout in the saturated state.

In the cycle *ABCH*, the condenser is only a little below the critical point, and it will be seen how small *HA* is compared with $H^1A^1$ in the second cycle, in which the critical temperature is not approached so close; and these lengths are proportional to the effective refrigeration in the two cases. The latent heat of evaporation vanishes at the critical point, and consequently so also does the corresponding change of entropy: the cooling effect would then become *nil*. With superheating, this result does not occur until the condenser temperature is 120° F. in the case examined above. These peculiarities in behaviour

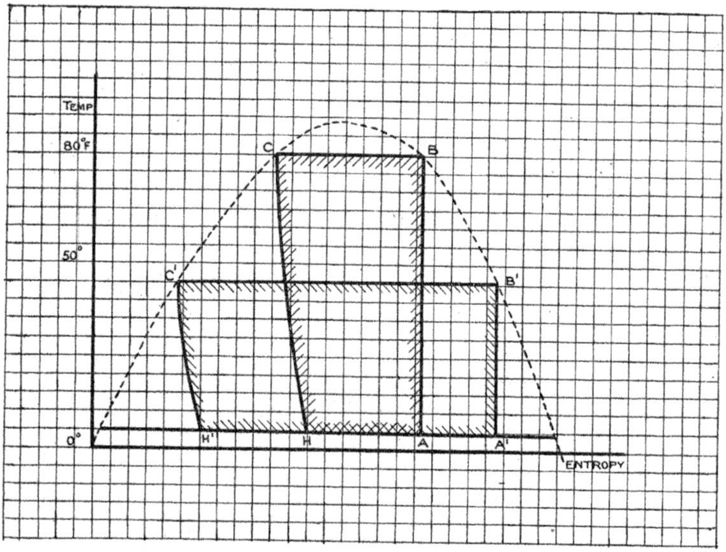


FIG. 17.

have no bearing on the performance of an ammonia machine, because its critical temperature (266° F.) is much beyond the range of actual practice.

The temperatures $T_1$ and $T_2$ are the temperatures of the working substance in the condenser and refrigerator respectively. These will always differ from the temperatures of the condenser water and the stores which are being kept cool. The former will be lower than $T_1$, and the latter higher than $T_2$; the effective lowering of temperature produced by the machine will always be less than $T_1 - T_2$. If we could work with this difference vanishingly small, the coefficient of performance would be thereby improved. The rate at which heat would leave the stores or be removed by the condenser water would then, however, vanish, since for its removal we depend upon conduction or radiation, which only take place down a gradient of temperature. This, therefore, is an inevitable source of loss in practice, and circumstances alone can decide as to how much is tolerable. Of course, where it might be required to *lower the temperature of a body quickly*, as in a sharp freezer (in contrast to the usual cases in which we merely require to keep it cool), considerable fall of temperature between the stores and refrigerator must needs be put up with, though at a corresponding sacrifice of economy.

# CHAPTER III.

## HISTORICAL.

THERE is little profit, from a practical point of view, in going deeply or minutely into the history of refrigerating or any other kind of machinery. From the present day standpoint a historical survey can only be of academic interest, and, therefore, a very brief reference to the early workers and early machines will be all that is called for in a volume such as this. As refrigerating machinery is now restricted to three, or at the most four types, each having its well-recognised advantages and disadvantages, it will be useful, before enumerating the early machines, to indicate the several systems upon which the various inventors exercised their talents.

These are clearly divided into three distinct classes, as follows:—

(1) The chemical refrigerator or freezing mixture;

(2) Machinery using a gas (air) which is compressed, partly cooled, and then reduced to a low temperature by expansion while doing work, the low temperature thus produced being employed to abstract heat from the substance it is desired to cool; and

(3) Machinery which produces its refrigerating effect by the evaporation of some volatile liquid having a low boiling-point.

## Class I.—The Freezing Mixture.

That a low temperature could be produced by the combination of two solid bodies uniting chemically to form a liquid has probably been known by scientists for centuries. In 1762 a mixture of crushed ice or snow and salt was employed by Fahrenheit when he obtained his zero temperature and placed the freezing-point of water at 32. The action of the frigorific mixture is due to the absorption of latent heat. When, for instance, certain solid substances, such as ice and salt, are brought together, the tendency to combine and melt is so great that heat absorbed in the action cannot be supplied from external sources as quickly as it is required by the two substances. Consequently, the internal heat of the two substances themselves is drawn upon, they become chilled, and the temperature drops until the heat drawn from the two sources—external and internal—is sufficient to keep pace with the rate of melting. Several attempts have been made to apply these properties on a commercial scale, and Sir William Siemens invented and patented a machine in which cold was produced by the heat absorbed by the fusion of calcium chloride and water. The machine was made, and produced ice, but was abandoned as expensive and impracticable. Tocelli invented a machine, using a mixture of ammonium nitrate and water, and the same agent has been used in other machines. Apart, however, from the popular objection to the use of chemicals in connection with ice-making, the machines were inefficient and cumbersome, and have long ago given place to mechanical methods. Freezing mixtures are, however, still used in the laboratory, where they are invaluable for quickly producing low temperatures for experimental purposes.

## CLASS II.—MACHINES USING GAS, OR COLD-AIR MACHINES.

Until about twelve years ago the cold-air machine was, at any rate in England and the Colonies, the most commercially successful method of producing low temperatures. The first cold-air machine of which there is any record, or which attained any success, was invented about the year 1845 by Dr John Gorrie of Apalachicola, Florida, U.S.A., and a monument has recently been erected in that town to his memory, as a pioneer of the refrigerating industry, by the members of the Southern Ice Exchange.

Gorrie's machines worked on what is known as a closed cycle. The air when cooled by expansion, instead of being discharged into the cold chamber as in the modern or open cycle machine, was used to cool brine, which in its turn was used to cool the cold chamber or to make ice, the expanded air being recompressed and used over again. The machine has been described as follows by Mr Coleman[1]:—

"The first machine consisted of a cylinder about 8 or 9 inches in diameter for compressing air, and of another cylinder about two-thirds the capacity of the first, in which the air was expanded. A jet of water was injected into the compressing cylinder, and the compressed air was discharged into a vessel surrounded by cold water, where it was further cooled and the injected water allowed to settle; the air was then admitted to the expansion cylinder, and, whilst expanding, a jet of salt water was injected. This salt water being drawn off, was allowed to circulate round vessels containing the fresh water to be frozen."

In 1861 Kirk's machine appeared and was fully de-

[1] *Proceedings Inst. C.E.*, vol. lxviii. p. 152.

scribed in a paper read before the Institution of Civil Engineers in 1874.[1] This machine, which also had a closed cycle like Gorrie's, was a practical success, and was used in several places, notably at Bathgate near Edinburgh, where it replaced one of Harrison's ether machines, and in Hong Kong, where it was used for making ice.

In 1868 Postle invented a cold-air machine, and in 1869 two German engineers, Windhausen and Nehrlick, invented machines for the same purpose. In Windhausen's machine the pressure of the expanded air was much below that of the atmosphere, and for this reason the machine was large and cumbersome. The necessity of maintaining a vacuum in the cold chambers was also a great disadvantage. Paul Giffard's machine was invented in 1873, and worked on the open cycle; the air, after expansion, being discharged direct into the chamber containing the articles to be cooled. At this time cold-air machines were receiving the bulk of the attention of inventors. In 1877 James Coleman and John and Henry Bell took the matter up, and the first Bell-Coleman machine was built in that year by the Bell-Coleman Mechanical Refrigeration Company of Glasgow. This machine aroused considerable commercial interest. With it the refrigerating industry may be said to have been fairly started, as it was used for preserving fresh meat in the passage across the Atlantic.

A somewhat similar machine was fitted on board the *Strathleven*, and in February 1879 she arrived in London with 34 tons of frozen mutton and beef on board. This was the first meat ever imported from Australia; for though about the year 1860 James Harrison had sent home a consignment of meat preserved by artificial ice,

[1] *Proceedings Inst. C.E.*, vol. xxxvii. p. 244.

the ice all melted before the end of the voyage, and the meat had to be thrown overboard.

Sir Montague Nelson, in a paper read before the Society of Arts in 1895, gives an interesting account of the difficulties experienced with these early shipments. In 1881 the sailing-ship *Dunedin* was fitted out with Bell-Coleman machinery by the New Zealand and Australian Land Company, and 5000 sheep were frozen and shipped in her. While in port no difficulties were met with. "Trouble, however, began," says Sir Montague Nelson, "when the vessel was becalmed in the tropics and the temperature was gradually found to be rising in the 'tween decks in spite of every effort on the part of the engineers to keep it below freezing-point. 'The hour brings the man,' and this time the man turned up in Captain Whitson, in command of the ship, who proved himself worthy of the grateful thanks which he afterwards received from all concerned in the venture, for, had he not been a man of ample courage and resource, there would have been, at any rate, a partially injured cargo to deal with in London. The main air trunk" (the trunk conveying the air from the expansion cylinder of the refrigerating machine to the hold) "got snowed up, and the Captain himself crawled down to clear it out and was successful, but got so benumbed during the process that he had to be hauled out by the heels with a rope. Captain Whitson's lot was not a happy one during this voyage, as in addition to the anxiety of maintaining cold below deck, he was, on certain tacks of the ship, in constant dread of setting the sails and rigging on fire by sparks from the refrigerator engine funnel. However, in the end all difficulties were overcome, and, after his strangely equipped craft had received much attention in the Channel from passing steamers, which offered towing assistance to what they could only imagine was a disabled steamer, he landed his cargo in splendid order."

The mutton fetched over 6d. per pound, and the possibilities of trade this opened up created considerable excitement in engineering and commercial quarters. About this time Messrs Haslam of Derby, Messrs J. & E. Hall of Dartford, and some other makers also started the manufacture of refrigerating machinery. In 1884 Messrs Haslam & Company, as the result of a legal action, bought up the Bell-Coleman patents, and eventually brought the manufacture of cold-air machines to a state of great perfection.

### Class III.—Machinery which produces its Refrigerating Effect by the Evaporation of some Volatile Liquid having a Low Boiling-point.

The earliest machine which produced ice by purely mechanical means came under this heading and was invented as far back as 1755 by William Cullen. It worked on what is known as the vacuum process, the evaporation of the water at the proper temperature being obtained by the use of a high vacuum maintained by an air-pump. Owing, however, to the high vapour tension of water, a very perfect vacuum was necessary in order to secure a sufficiently intense refrigeration, and the mechanical difficulty of this led to the introduction, in 1810, of sulphuric acid as an absorbing agent for the water and vapour. This acid, working in conjunction with the vacuum, enabled ice to be made with comparative ease. Cullen's machine, modified, was probably the basis of Vallance's sulphuric acid machine invented in

1824, and of which Edmund Carré's machine, invented about 1850, was an improvement. Carré's machine consisted of a glass vessel containing water, which, under the influence of vacuum, and in connection with a supply of sulphuric acid, froze rapidly into a soft spongy mass. The vacuum was produced by a hand air-pump, and the water was frozen in about five minutes from starting operations. The machine, when invented, was much used in Paris, and an improved design is still manufactured for domestic purposes. The necessity, however, of renewing the sulphuric acid from time to time has probably prevented its more universal adoption, as there is great need, especially in out of the way districts, for a handy domestic ice-maker. A few years later Edmund Carré's brother Ferdinand invented an ammonia-absorption machine, which was the forerunner of the absorption machines of the present time. Ferdinand Carré's first machine was exceedingly crude, and consisted only of two vessels—one used as a generator, and the other as an absorber. The generator contained the solution of ammonia and water, and, on being heated, the ammonia gas was driven off to the refrigerator, where it was liquefied by water applied externally. On the solution in the generator becoming sufficiently dilute, the heater was removed and the solution was cooled by water, when the liquefied gas in the refrigerator at once began to volatilize, and to be reabsorbed by the cooled dilute solution. In evaporating it absorbed heat from the water to be frozen, which was placed in a recess in the refrigerator. The machine was afterwards vastly improved by inventors in France and Germany, by Reece and Stanley in England, and by Mort and Nicolle in Australia. The improvements consisted of the addition of new vessels and the development of the process from an intermittent one, as designed by Carré, to a continuous one. Reece, in 1876, introduced several notable improvements, particularly the analyzer for rendering the ammonia gas anhydrous, and the absorber, in which the gas, after evaporation, was reabsorbed by the dilute solution or weak liquor. In 1870 Mort introduced the interchanger, or economizer, designed to transfer heat from the weak solution leaving the generator to the strong solution entering it from the absorber. Later on, Stanley introduced steam-coils into the generator.

To Eugene Nicolle, working in collaboration with Mort and Morris, belongs the credit, it is believed, of making, in 1867, the first definite designs for the installation of a refrigerating machine on board ship for the purpose of bringing frozen meat from Australia to England; and to Pontifex and Wood belongs the credit of first bringing the absorption machine to a high state of perfection and a considerable state of efficiency. Their designs, greatly improved, are in use at the present day by Messrs Haslam and Company, of Derby.

To return for a moment to the vacuum machine. In 1870 Windhausen, who also invented the cold-air machine previously referred to, patented a vacuum machine, which was to be capable of producing ice direct from the water without the assistance of sulphuric acid, but which was arranged in such a way that sulphuric acid could, if necessary, be also applied. In 1881, after several improvements had been effected, a machine was built and erected at the Aylesbury Dairy Company's premises at Bayswater, but was afterwards removed to the Lillie Road, Earl's Court. The machine was designed to produce from 12 to 15 tons of ice per day, and attracted considerable attention, papers describing it being read before the Society of Engineers and the Society of Arts.[1] The machine has been described as follows[2]:—

"It consists of six slightly tapered ice-forming vessels of cast-iron, of circular cross-section, closed at their bottom ends by hinged doors with air-tight joints, into which water is allowed to flow at a regular rate through suitable nozzles, the cylinders being steam-jacketed in order to allow the ice to be readily discharged. The upper parts of these vessels communicate with the pump through a long horizontal iron vessel of circular section containing sulphuric acid, which, when the machine is in operation, is kept in continual agitation by means of revolving arms. The acid vessel is surrounded with cold water, which carries off most of the heat liberated during the absorption of the vapour. The pump has two cylinders, one double-acting of large size, and a smaller single-acting one. The capacities of these cylinders per revolution are as 62 to 1. The air and whatever vapour has passed the acid are drawn into the large pump, which partially compresses and delivers them into a condenser. Here part of the vapour is condensed by the action of cold water, the remainder passing along with the air to the second pump, where they are compressed up to atmospheric tension and discharged. The advantage gained by the use of a compound pump is due to the action of the intermediate condenser and to the compression being performed in two stages, by which the losses from the clearance spaces in the large pump are rendered much less than they would be if compression to atmospheric

[1] *Transactions of the Society of Engineers*, 1882, p. 145; *Journal of the Society of Arts*, 1882, vol. xxxi. p. 20.

[2] *Proceedings of the I. Mech. E.*, 1886, No. 2, p. 207.

pressure were accomplished in a single operation. The effect of the pump is said to be such that a vacuum of half a millimetre of mercury, or about 0·0097 lb. per square inch, can be continuously maintained, though in actual work about 2½ millimetres, or 0·0484 lb. per square inch, is as low as is necessary. The concentration of the acid is effected in a lead-lined vessel, in which is a coil of lead piping heated by steam, the pressure in the vessel being kept down by means of an ordinary air-pump. No acid pump is needed, as the transfer from one vessel to another is effected by the pressure of the atmosphere. The comparatively cool weak acid, on its way to the concentrator, is heated in an interchanger by the strong acid returning from the concentrator. Six blocks of ice, each weighing about 560 lbs., are formed in about sixty minutes after starting. The charge of acid is said to serve for four makings of ice, after which it becomes too weak and requires to be concentrated."

The great fault of the machine was that the ice produced was opaque and porous, or what is known as rotten, and rapidly melted when exposed to the atmosphere. This objection is inherent to the system, and is caused by the fact that the water, being admitted in the form of spray, freezes into a series of minute globules, which adhere together, but which contain air in and between them. Several attempts have been made to overcome this difficulty, but no great measure of success has yet been obtained. In 1878 Harrison invented a vacuum machine, which, though ingenious, does not call for any special mention, and was, it is believed, never used for commercial purposes. Recently, attempts have been made to revive the vacuum and sulphuric process in America. Water is introduced into a cylinder, which is under a vacuum and in communication with an "absorbing agent" (sulphuric acid). The water is frozen solid, and is afterwards cut up into convenient pieces by a circular saw. The ice is opaque, and special advantages are claimed for it on the ground of its whiteness. In England opaque ice is, except for the fish trade and similar purposes where it is only wanted to be cold, almost unsaleable, and the vacuum process has practically disappeared.

## The Compression Process.

What may be regarded as the forerunner of the modern compression machine was invented in 1834 by Jacob Perkins. The machine has been described by Sir Frederick Bramwell,[1] and has since been often illustrated and described in various papers and journals. However, as it occupies such an important place in the history of the science, a further brief description will not be out of place.

Perkins intended the machine to work with sulphuric ether, but, as a matter of fact, the machine never got beyond the experimental stages. During the experiments caoutchoucine, a volatile oil produced by the distillation of caoutchouc, or india-rubber, at a high temperature, was the agent used. In the machine itself there are all the essential parts of the compression machine of to-day. The water to be frozen was contained in a vessel which was jacketed by a second vessel containing the refrigerating agent, and ingress of external heat was prevented by surrounding the two vessels with insulating materials. The gas or vapour caused by the evaporation of the refrigerating agent was drawn off by the compression pump, and was compressed and delivered to the coils

[1] *Journal of the Society of Arts*, 1882, vol. xxxi. p. 77.

forming the condenser. The pressure was sufficient, with the aid of the water circulating round the coils, to cause liquefaction, and the liquid ran off through a valve to be re-evaporated in the jacket of the refrigerator. It was intended at one time to use these machines in an ice factory, but the project fell through, and it was not until twenty-three years later that James Harrison of Geelong, N.S.W., improved the compression machine, and, with sulphuric ether as an agent, brought it into practical use in breweries as well as in the refrigeration of meat and other perishable products. Its manufacture in New South Wales was taken up by Messrs Russell & Company of Sydney, and in England by Messrs Siebe & Gorman. It should be a matter of congratulation and pride to all Australasians that the first really practical compression machine was invented by a New South Wales engineer, who thus paved the way for the enormous developments that have taken place in Australasia due to the frozen meat industry. It has been noted that Dr Harrison sent the first consignment of meat to this country preserved by artificial ice made by his machine, and it is to be hoped that in due time his labours and genius will be honoured by the erection of some similar memorial as has recently been put up by the Southern Ice Exchange to Dr Gorrie in the United States. As a refrigerating agent ether is now almost obsolete. It has many disadvantages, prominent among them being its inflammability. The size of the machine for a given duty was also very large compared with one using ammonia or $CO_2$, and, as both the forward and back pressures were below that of the atmosphere, difficulty used to be experienced in keeping air from leaking in at the compressor glands. In 1873 Professor Linde introduced the ammonia-compression machine, which is now so well known. Three years later, a machine, working on the compression system and using sulphur dioxide as a refrigerating agent, was introduced by Raoul Pictet of Geneva, and is still in use to a considerable extent abroad. Pictet discarded lubrication in the compressor, and was probably the first to use a water-jacket to absorb some of the heat of compression, not only the compressor itself, but also the piston and piston-rod, being cooled in this way.

The seventies, more particularly in the United States, were pregnant with patents and improvements in refrigerating machinery. It would require too much space to describe the various men and machines at length, but, to select a few out of many, mention must be made of David Boyle, whose career has been ably described by Mr Skinkle, and who is said to have perfected his designs during the siege of Vicksburg; Charles Ball, who considerably developed the absorption machine; Ferguson, who designed the first machines for the Frick Company; and John De la Vergne, who introduced oil-injection, and whose name is now world-known.

## The Binary System.

Yet another system of refrigeration remains to be described. This is known as the Binary or Double Agent System, and the refrigerating effect is brought about either by the use of a liquid, being a compound of two agents, one of which will liquefy at a low pressure and then absorb the other, or by the use of a simple agent which is liquefied partly by compression and partly by absorption.

The principal agent used in the former system is a mixture of ether and sulphur dioxide, it being found, after numerous experiments with other mixtures, that this combination was the most suitable, as the ether possesses at ordinary temperatures an absorbing power for sulphur dioxide of some 300 times its own volume, while at 60° F. the vapour tension of the compound was below atmospheric pressure. A machine using this compound has been used in the United States by De Mota and Rossi.

A machine using a cycle of the second class was introduced in 1869 in Sydney, N.S.W., by Mort and Nicolle, who have been previously referred to. In this ammonia was used with water as an absorbent. The refrigerator contained a strong solution of ammonia and water, and a pump was used to reduce the pressure in this vessel, and so vaporize the ammonia. The gas discharged from the pump was brought into contact with the weak liquor remaining, and was again absorbed, the resulting strong liquor being once more forced to the refrigerator. The machine, however, did not come into general use.

About the year 1884 Pictet, the inventor of the sulphur-dioxide machine, invented a compound agent which is universally known as Pictet's liquid. It was composed of certain proportions of carbon dioxide ($CO_2$) and sulphur dioxide ($SO_2$), the complete formula for the mixture being $CO_4S$. The liquid resulting from this combination has a vapour tension which is much less than $CO_2$ at all temperatures, and for temperatures above 78° F. is below sulphur dioxide itself. The liquid, though it attracted considerable attention at the time of its discovery, has not come into general use as a refrigerating agent.

Quite a number of agents have from time to time been proposed, and some used experimentally for refrigerating machines. Among these may be mentioned chloride of methyl, ethyl chloride, ethylic ether, ethylo-sulphurous dioxide (Mota and Rossi's mixture), and an obscure agent called cryogene. Recently acetylene ($C_2H_2$) has been proposed as a refrigerating agent; it is said to require a pressure of 720 pounds on the square inch to produce liquefaction at 32° F. Its intense inflammability would certainly militate against it. Other agents, such as naphtha, gasoline, etc., have been experimented with, more particularly in the United States, but there, as in England and elsewhere, the only two agents which have any extended use and which require consideration are anhydrous ammonia ($NH_3$) and carbon dioxide ($CO_2$).

# CHAPTER IV.

## THE PROPERTIES OF AMMONIA AND $CO_2$.

For the convenience of comparison the critical data of the three principal refrigerating agents were given in the Introduction. It will, however, perhaps be useful to repeat those for ammonia as follows:—

| Pressure in lbs. per sq. in. at 0° F. | Heat of vaporization per lb. at 0° F. | Volume cub. ft. per lb. at zero F. | Specific heat of liquid. | Heat of vaporization per cub. ft. at 0° F. |
|---|---|---|---|---|
| 30 | 555·5 | 9·10 | 1·02 | 61·7 |

Ammonia, the agent which, with the exception of carbon dioxide ($CO_2$) is now almost exclusively used in refrigerating machines, is a combination of nitrogen and hydrogen corresponding to the formula $NH_3$—that is to say, one part of nitrogen in combination with three parts of hydrogen, or, if by weight, 82·4 per cent. of nitrogen and 17·6 per cent. of hydrogen. It is an alkali, and in its natural state is a gas. At the pressure of the atmosphere it will not liquefy until it is reduced to a temperature of −30° F. Its liquefaction at higher temperatures is, in accordance with the law of gases, merely a question of pressure.

Anhydrous ammonia, when pure, is a colourless liquid with a peculiar caustic taste. It is manufactured from sulphate of ammonia treated with lime and from other bye-products, but chiefly from the ammoniacal liquor given off in the preparation of coal gas. It is also prepared from aqua ammonia of 26° Beaumé, or containing 38½ per cent. of ammonia. In the early days it was made directly from its mineral salts, and was absolutely anhydrous. The price of the liquid manufactured in this way is, however, prohibitive.

The process of manufacture is very simple, and is generally similar to the cycle employed by the absorption machine, which, indeed, is sometimes used for the purpose.

The operation is as follows:—

The material from which the anhydrous ammonia is to be manufactured is placed in a cast-iron vessel termed a still, which is fitted with a steam-coil. These are marked respectively *A* and *B* in the diagram (fig. 18). Steam is admitted to the coil, and the heat drives off the $NH_3$, leaving the residue behind. The ammonia escapes through the pipe at the top of the still into the vessel *C*. This is called the separator, and is filled with cast-iron plates. It somewhat resembles the analyzer of the absorption machine, to which, indeed, its function is similar. Any water carried over with the ammonia gas is caught by these plates and falls back into the still *A*. The ammonia gas proceeds onwards to a condenser worm marked *D*. This is surrounded by water, which condenses the gas and also any watery vapour that may have escaped as steam past the cast-iron trays of the separator. The mixture of liquid ammonia and water is expanded into two vessels *E* and *F*, commonly called Wolffs' bottles, where it turns into gas and is further purified. These two vessels are placed in series, and the gas passes through them one after the other. From the top of *F* the gas passes to a vessel, *G*, known as the dehydrator. In this vessel

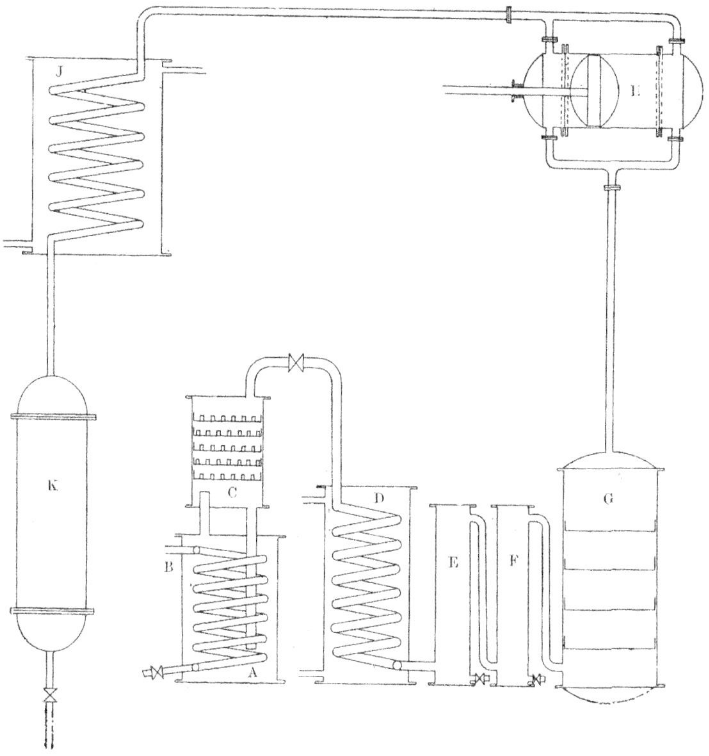


FIG. 18.

trays of quick-lime are placed, and, as the gas is drawn through and over the lime by the suction of the compressor *H*, every particle of water it contains is absorbed, and it passes away in a perfectly anhydrous condition to the compressor. Here it is compressed in the usual way and delivered to the ammonia condenser *J*, which can, of course, be either of the open air or of the submerged type, which latter, for convenience sake, has been shown. There it is condensed by the cooling action of the water, and in liquid form flows down into the receiver *K*. From this vessel the ammonia flasks are filled for sale.

The Wolffs' bottles are fitted with drain cocks, through which any impurities which have passed over with the liquid ammonia are drawn off. The burnt lime gradually loses its power, as it absorbs moisture and has to be removed from time to time. Sometimes caustic soda is used instead of lime. Lime when slacked assumes a powdery form, and is apt to choke the dehydrator, but, needless to say, it should be renewed before this happens. Caustic soda is an exceedingly unpleasant material with which to have to deal, and for this reason its use has been largely discontinued. The form of still, used, varies with the nature of the material from which the $NH_3$ is being manufactured. Some stills are made capable of standing an internal pressure of some pounds on the square inch. The object of this is to enable the strong liquor in the still to be raised above the temperature of boiling-point—212° F.—without ebullition, thus partly preventing evaporation of the water, which is the main object in distilling $NH_3$.

Ammonia is stated to have the distinction of being the only volatile alkali known to science. Its critical temperature is 266° F., and at a temperature of 1600° [1] F. it

[1] 1600° F. is the temperature of practically complete dissociation. The process starts at a lower temperature (about 900° F.) and a gradual change takes place, a definite fraction splitting up at each temperature.

will dissociate into its constituent gases. At a temperature of $-94°$ F. it freezes solid. All these temperatures are beyond the normal range met with in practice, and the refrigerating engineer need not feel perturbed that his plant will be overtaken by catastrophes traceable to the action of these characteristics. Of course, ammonia for use in a compression system should be perfectly free from water and other impurities of any sort, one-tenth of one per cent. being all that should be allowed.

A curious feature of modern ammonia is that, though when tested it will show itself to be quite pure within the above limit and will give no trace of water, it will in some way or other produce impurities after it has been at work for some time in the plant. For this reason no modern compression plant should be erected without a dryer such as is described on page 146.

Testing ammonia for impurities requires to be done with great care. Being an unpleasant material to handle, it should not be undertaken without previous experience, and it is best to let a properly qualified analyst with proper appliances at his command do the work when possible. For rough tests the usual method is partially to fill a glass test-tube with liquid drawn from the system. This test-tube should have a cork fitted with a vent-pipe to carry off the gas as it evaporates, the other end of the vent-pipe being immersed in water which will absorb the gas. If the liquid is pure it will all boil away, leaving no residue, but if it contains impurities these will be left behind in the test-tube.

The following directions[1] for testing ammonia for water or other impurities may also be observed.

As shown by the engraving, screw into the ammonia

[1] Messrs the Frick Company, Waynesboro, Pa., U.S.A.

flask a piece of bent $\frac{1}{4}$-in. pipe, so that a small test-bottle may be placed in a position to receive the discharge from it. This test-bottle should be of thin glass with a wide neck, so that the $\frac{1}{4}$-inch pipe can pass readily into it, and of about 200 cubic centimetres capacity. Put the wrench on the valve, and tap it gently with a hammer. Fill the bottle about one-third full, and throw the sample away in order to purge the valve, pipe, and bottle. Quickly wipe off the moisture that has accumulated on the pipe,

FIG. 19.

replace the bottle, and open valve gently, filling the bottle about half full. This last operation should not occupy more than one minute. Remove the bottle at once, and insert in its neck a stopper with a vent-hole for the escape of the gas. A rubber stopper with a glass tube in it is the best, but a rough wooden stopper loosely put in will answer the purpose. Procure a piece of solid iron that should weigh not less than eight or ten pounds, pour a little water on this, and place the bottle on the wet place. The ammonia will at once begin to boil, and in warm weather will soon evaporate. If there is any residuum, pour it out gently, counting the drops carefully. Eighteen drops are about equal to one cubic centimetre, and, if the sample taken amounted to 100 cubic centimetres, the percentage of the liquid remaining can be readily approximated.

Care is necessary in drawing off the sample, as a very little moisture in the bottle or in the pipe, or a brief exposure to the atmosphere, will at once affect it.

The liquid may also be drawn from the liquid receiver, and instead of the 200 c.c. capacity bottle, it can equally well be drawn off into a graduated test-tube provided with a cork and a glass tube, which should however be bent so that it can be immersed in water which will absorb the ammonia gas as it escapes.

A great deal of nonsense has from time to time been talked about the danger of an escape of ammonia causing a fire. The folly of this will be realized when it is considered that neither ammonia itself nor nitrogen, of which it contains 82·4 per cent., can promote or support combustion. Where explosions or conflagrations have followed an explosion or escape of ammonia, it has in every case been due to the shower of finely divided mineral lubricating oil, which has been blown out of the system with the gas. Ammonia itself is an extinguisher of fire, and aqua ammonia sometimes forms the mysterious liquid to be found in the interior of patent fire-extinguishers.

As ammonia is so readily absorbed by water, a serious leak may exist and yet not be discovered. It is advisable, therefore, from time to time to test the water used for condensing purposes, or the brine surrounding the expansion coils, with Nessler's reagent,[1] the prescription for which is as follows:—

[1] J. E. Siebel, *Compend of Mechanical Refrigeration.*

"Dissolve 17 grams of mercuric chloride in about 300 c.c. of distilled water; dissolve 35 grams of potassium iodide in 100 c.c. of water; add the former solution to the latter, with constant stirring, until a slight permanent red precipitate is produced. Next dissolve 120 grams of potassium hydrate in about 200 c.c. of water; allow the solution to cool; add it to the above solution, and make up with water to one litre; then add mercuric chloride solution until a permanent precipitate again forms; allow to stand till settled, and decant off the clear solution for use; keep it in glass-stoppered blue bottles, and set away in a dark place to keep it from decomposing."

This reagent can readily be obtained from any chemist after the above prescription, and all that is required is to add a few drops to a test-glass full of the suspected water or brine, which will assume a yellow colour if ammonia is present at all, or a full brown if it is present in any quantity.

The curve of vapour tensions of anhydrous ammonia, showing graphically the pressure of the gas in pounds per square inch for various temperatures, is given in fig. 20.[1] Its latent heat of vaporization for various temperatures is shown in the table at the end of the book, and it will be noticed that at atmospheric pressure its temperature is −30° F., and its latent heat of vaporization—that is to say, the amount of heat that it will have to absorb before it will turn from a liquid to a gas—is 573·69. At this pressure one pound of the gas has a volume of 18·693, and the liquid a volume of ·02374; while the weight of the vapour is ·0535, and of the liquid 42·123. The

[1] In the vertical column of pounds per square inch the bottom figure should be 10.

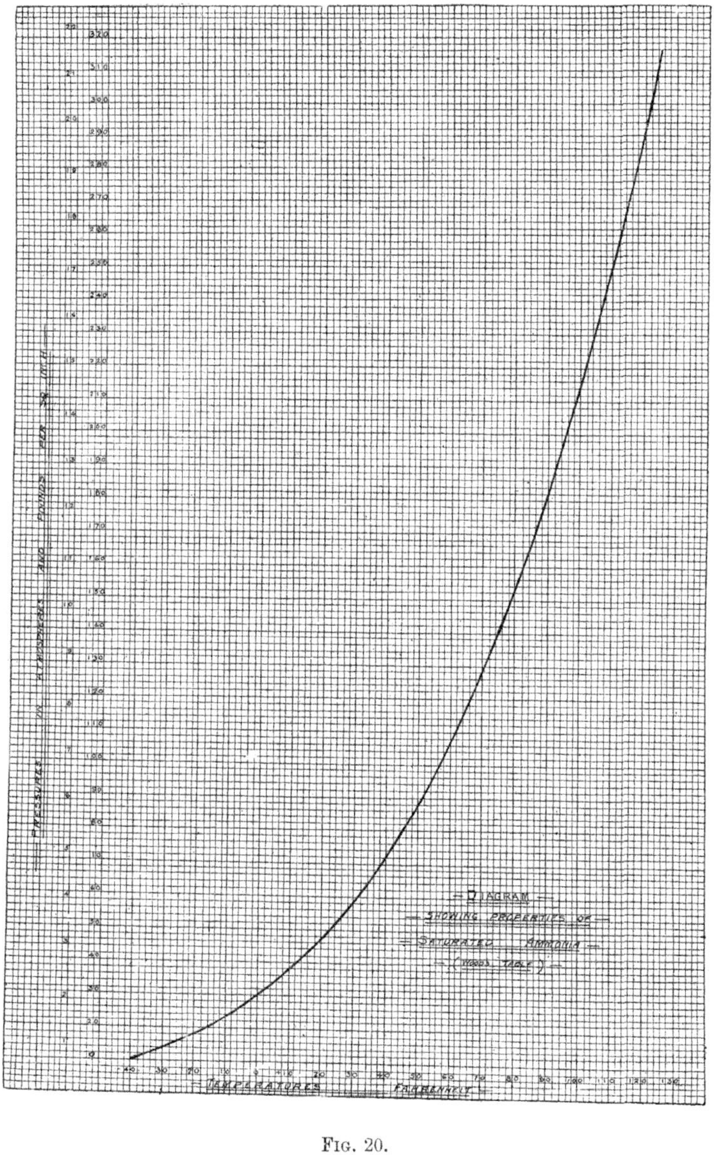
DIAGRAM
SHOWING PROPERTIES OF
SATURATED AMMONIA
(WOODS TABLE)
PRESSURES IN ATMOSPHERES AND POUNDS PER SQ. INCH
TEMPERATURES FAHRENHEIT

Fig. 20.

properties of the gas for other pressures and temperatures can readily be found from the table.

A glance at the curve will show how important it is from an economical standpoint to keep the pressure of the gas in the condenser as low as possible, and as this pressure is determined by the temperature of the condenser, it is essential for economical working to have ample cooling water at a low temperature. It will also be noticed that as the pressure of the gas falls the volume increases. At a pressure of 30 lbs. on the square inch the temperature is zero F. and the volume 9·028; at a pressure of 14·13 lbs. on the square inch the temperature is $-30°$ F. and the volume 18·693, or more than double. This means that, owing to the increased volume of the gas, the compressor has, at the lower temperature, to take twice the number of strokes per minute, or, if the speed is kept constant, to be twice the size, in order to pass the same quantity of gas and do the same amount of work as it would were the higher temperature in use. For this reason, and from this point of view, it is preferable, where possible, to expand the ammonia gas directly into pipes placed in the cold rooms, or liquid to be cooled, instead of effecting a transmission of heat by first cooling and then circulating brine; there are, however, cases where this cannot be done, but of these more hereafter.

**Ammonia Accidents—First Aid.**—Ammonia is incapable of being respired, and will produce suffocation if inhaled in any quantity.

Several fatal accidents have at times been caused by the bursting of portions of an ammonia plant, and the consequent escape of large volumes of the gas into the rooms where workmen are employed. In emergencies of this sort the first action is naturally to extract any person who may be overcome from the room with the least delay possible. If a sponge, towel, or lump of waste is saturated with water and held over the mouth and nose, a room charged with ammonia gas can be entered and the person overcome rescued. The first thing to do is, obviously, to send for the nearest doctor, but life may often be saved and pain lessened by at once administering a glass of beer or a dose of common vinegar or very weak acetic acid. Either of these will act as an antidote to the ammonia.

## Carbonic Anhydride.

This gas, commonly known by its chemical symbol of $CO_2$, is probably the most familiar of all the gases. It is responsible for the sparkle of champagne, beer, and aerated waters, and exists in nearly all water. In its natural state it is a colourless, odourless gas, and, at a temperature of 32° F. and a pressure of 36 atmospheres, it can be converted into a colourless, odourless liquid.

When carbon is burned in oxygen, $CO_2$ is formed, but if the supply of oxygen is insufficient, carbonic oxide (CO), an extremely poisonous gas, is also produced. The latter will burn with a blue flame, and by its combustion produce $CO_2$, but pure $CO_2$ will neither originate nor support combustion. $CO_2$ does not possess the affinity for water that is common to ammonia, but it will dissolve in about its own volume, forming carbonic acid ($H_2CO_3$). The solution has a sharp acid taste, and will turn blue litmus paper red. The volume of $CO_2$ that water is capable of absorbing decreases as the temperature rises, as pointed out on page 167 (water-softening). All $CO_2$ can be expelled if the water is brought to boiling-point.

**The Properties of $CO_2$.**—As a refrigerant $CO_2$ is very

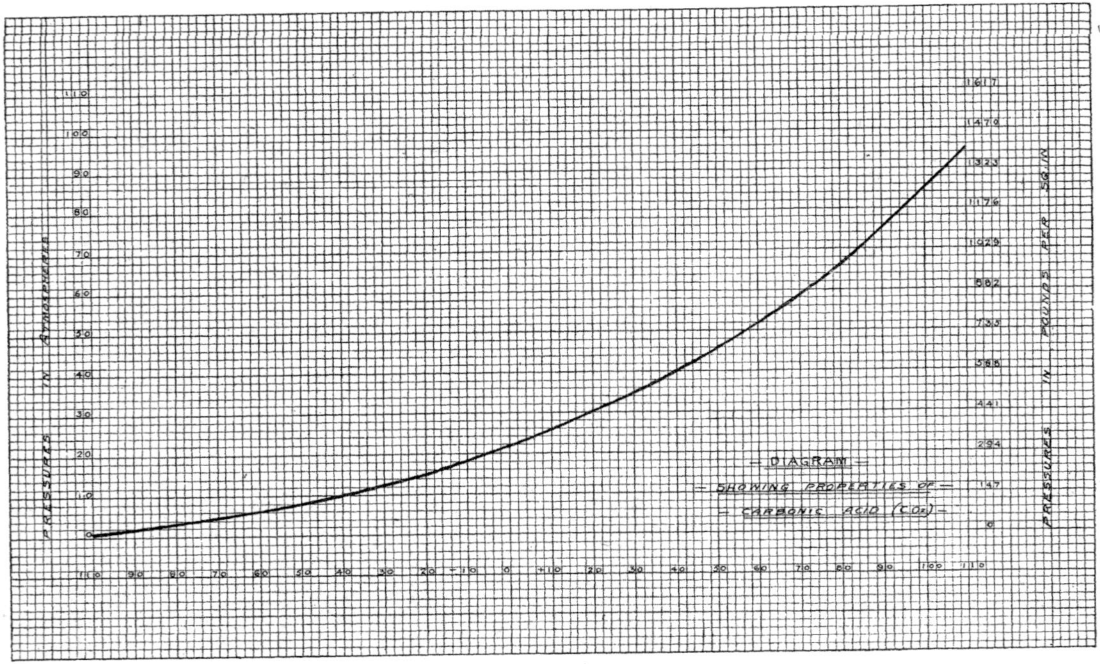
PRESSURES IN ATMOSPHERES
110
100
90
80
70
60
50
40
30
20
10
0
110 90 80 70 60 50 40 30 20 −10 0 +10 20 30 40 50 60 70 80 90 100 110
1617
1470
1323
1176
1029
882
735
588
441
294
147
0
PRESSURES IN POUNDS PER SQ. IN
— DIAGRAM —
— SHOWING PROPERTIES OF —
— CARBONIC ACID (CO₂) —

FIG. 21.

## PROPERTIES OF SATURATED CARBONIC ACID GAS. (SCHRÖTER.)

| Temperature of ebullition in degs. F. | Absolute pressure in lbs. per sq. in. | Total heat reckoned from 32° F. | Heat of liquid reckoned from 32° F. | Latent heat of evaporation. | Heat equivalent of external work. | Internal latent heat. | Increase of volume during evaporation. | Density of vapour or weight of 1 cub. ft. |
|---|---|---|---|---|---|---|---|---|
| $t$ | $P \div 144$ | $\lambda$ | $q$ | $r$ | $APu$ | $\rho$ | $u$ | |
| (1) | (2) | (3) | (4) | (5) | (6) | (7) | (8) | (9) |
| – 22 | 210 | 98·35 | – 37·80 | 136·15 | 16·20 | 119·95 | ·4138 | 2·321 |
| – 13 | 249 | 99·14 | – 32·51 | 131·65 | 16·04 | 115·61 | ·3459 | 2·759 |
| – 4 | 292 | 99·88 | – 26·91 | 126·79 | 15·80 | 110·99 | ·2901 | 3·265 |
| 5 | 342 | 100·58 | – 20·92 | 121·50 | 15·50 | 106·00 | ·2435 | 3·853 |
| 14 | 396 | 101·21 | – 14·49 | 115·70 | 15·08 | 100·62 | ·2042 | 4·535 |
| 23 | 457 | 101·81 | – 7·56 | 109·37 | 14·58 | 94·79 | ·1711 | 5·331 |
| 32 | 525 | 102·35 | 0·00 | 102·35 | 13·93 | 88·42 | ·1426 | 6·265 |
| 41 | 599 | 102·84 | 8·32 | 94·52 | 13·14 | 81·38 | ·1177 | 7·374 |
| 50 | 680 | 103·24 | 17·60 | 85·64 | 12·15 | 73·49 | ·0960 | 8·708 |
| 59 | 768 | 103·59 | 28·22 | 75·37 | 10·91 | 64·46 | ·0763 | 10·356 |
| 68 | 864 | 103·84 | 40·86 | 62·98 | 9·29 | 53·69 | ·0577 | 12·480 |
| 77 | 968 | 103·95 | 57·06 | 46·89 | 7·06 | 39·83 | ·0391 | 15·475 |
| 86 | 1080 | 103·72 | 84·44 | 19·28 | 2·95 | 16·33 | ·0147 | 21·519 |

PROPERTIES OF SATURATED CARBONIC ACID VAPOUR. (TRANSFORMED FROM MOLLIER'S TABLE INTO BRITISH UNITS.)

| Pressure in lbs. per sq. in. | Temperature (Fahr.). | Specific volume of liquid in cub. ft. per lb. | Specific volume of saturated vapour in cub. ft. per lb. | Expansion on vaporization. | Latent heat in B.T.U. | Heat equivalent of external work. | Internal latent heat. |
|---|---|---|---|---|---|---|---|
| $P$ | | $v'$ | $v''$ | $v'' - v' = u$ | $r$ | $APu$ | $\rho$ |
| 284 | −5 | ·0161 | ·3182 | ·3021 | 119·86 | 15·98 | 103·88 |
| 355 | +9 | ·0165 | ·2497 | ·2332 | 113·26 | 15·41 | 97·85 |
| 427 | +20·7 | ·0170 | ·2028 | ·1858 | 106·94 | 14·76 | 92·18 |
| 498 | 31·2 | ·0176 | ·1698 | ·1522 | 100·44 | 14·06 | 86·38 |
| 569 | 40·5 | ·0181 | ·1435 | ·1254 | 93·76 | 13·26 | 80·50 |
| 640 | 49·0 | ·0188 | ·1232 | ·1044 | 86·90 | 12·42 | 74·48 |
| 711 | 56·7 | ·0194 | ·1059 | ·0865 | 79·65 | 11·45 | 68·10 |
| 782 | 63·8 | ·0204 | ·0919 | ·0715 | 71·73 | 10·39 | 61·34 |
| 853 | 70·3 | ·0213 | ·0795 | ·0582 | 63·22 | 9·22 | 54·00 |
| 924 | 76·7 | ·0226 | ·0676 | ·0450 | 52·87 | 7·74 | 45·13 |
| 995 | 82·6 | ·0249 | ·0563 | ·0314 | 39·29 | 5·79 | 33·50 |
| 1066 | 88·2 | ·0307 | ·0385 | ·0078 | 10·51 | 1·56 | 8·95 |

widely used. There was for some years considerable doubt as to its exact properties, Professor Schröter's figures as given in the table on page 91, and the corresponding temperatures and pressures as shown graphically in the curve fig. 21, being, however, generally accepted. Recently Dr Mollier, as a result of experiments, has published[1] a table giving different, and it is believed more correct, values than Professor Schröter. In this table the units have been directly altered to British units, as, considering the labour it would have involved, it was not thought worth while to change these by interpolation so as to give values in integral numbers of five every 50 lbs. per square inch. Throughout this work Dr Mollier's figures have been taken in preference to Professor Schröter's.

**The Manufacture of $CO_2$.**—The manufacture of $CO_2$ is carried on in several ways, and a large quantity of the pure liquid is obtained from the fermenting tuns of breweries. The fact that it is heavier than air makes collection in this case easy, and plants both in England and Germany have been erected for this purpose. It can also be readily obtained from other sources.

At the Carbonic Acid Gas Company's works[2] the method of manufacture employed is as follows:—The $CO_2$ is produced by burning coke in the furnaces of boilers. The gases from these furnaces, after being scrubbed, are brought into contact with potash lye, which possesses a strong affinity for $CO_2$, and absorbs it.

In order to remove the sulphur which is contained in the coke, and which, were it allowed to come into contact with

[1] R. Mollier, Ueber die kalorischen Eigenschaften der Kohlensäure ausserhalb des Sättigungsgebietes, *Zeitschrift für die gesammte Kälte-Industrie*, vol. iii., 1896.

[2] *Engineering*, Oct. 20, 1899.

the potash, would form potassium sulphate and render it useless as an absorber of $CO_2$, the furnace gases are scrubbed on their way to the potash towers by being passed through two towers placed in series and filled with broken marble or limestone. In the first the gas comes in contact with a stream of water which has been heated in an interchanger by the hot $CO_2$ passing from the boiler. In the second tower the gas comes in contact with a stream of cold water, and then passes to the absorbing towers, which are filled with trays containing broken coke. In these it meets a constant stream of potash lye, which, as before stated, absorbs the $CO_2$ and allows the rest of the gases to go to waste. The lye containing the $CO_2$ is pumped to the boiler, and on its way passes through an interchanger, where it absorbs heat from the regenerated lye, which, having parted with its $CO_2$, has just left them.

The boiler pressure is 5 lbs. per square inch, and the lye is passed through the boilers in series, the $CO_2$ passing off at the top through a liquid trap, which catches any moisture carried over and returns it to the regenerated lye storage tank. As the gas comes from the boilers, it is first passed through an interchanger, where it gives up its heat to the water supply to the first scrubbing-tower, and is then collected and passed to the compressors. These deal with the gas in three stages. In the first it is compressed to 25 lbs. per square inch, at which pressure the gas is further purified and robbed of any moisture it may have contained by being brought in contact with calcium chloride. The second compressor raises the pressure to 125 lbs. per square inch. Before entering the third and last compressor, the gas is cooled; it is then compressed to about 1000 lbs. per square inch, the exact pressure depending on the temperature of the condensing water; is liquefied and finally bottled.

**Cylinders for $NH_3$ or $CO_2$.**—Owing to the occurrence of several explosions of cylinders containing compressed gas, a committee was appointed by the Government in 1895 to inquire into the whole matter. The committee duly presented an exhaustive and interesting report,[1] its recommendations as regards the above being as follows:—

After setting forth the hydraulic and annealing tests, etc., to which the cylinders are to be subjected, it is stated that for $CO_2$ the maximum working pressure is to be reckoned as 120 atmospheres (1800 lbs. per square inch). All the cylinders are to be either of lap-welded wrought-iron or steel, or of seamless steel, and each cylinder is to be clearly stamped with the tare of the cylinder and the greatest permissible weight of $CO_2$ it may contain. This latter is limited to $\frac{3}{4}$ lb. of $CO_2$ for each pound of water capacity for this country, and $\frac{2}{3}$ lb. of $CO_2$ per lb. of water capacity for the tropics.

Before being sent out, the cylinders are to be labelled, stating that they contain $CO_2$, and that they are to be kept in a cool place. If the cylinders are sent out unpacked, the valves are to be protected by a metal cap.

For $NH_3$ the same general recommendations apply, but the maximum working pressure is fixed at 1000 lbs. per square inch, and each cylinder is, before being used, to be tested hydraulically to 1500 lbs. per square inch. The cylinders are not to contain more than $\frac{1}{2}$ lb. of $NH_3$ per lb. of water capacity.

Generally it is recommended that each cylinder should be marked with a rotation number, a maker's or owner's

[1] Parliamentary Papers C 7952—Committee on the Manufacture of Compressed Gas Cylinders.

mark, and an annealing and test mark with date, and that tests should be repeated at intervals of two years, and annealing at intervals of four years.

With $CO_2$ cylinders it is the practice of Messrs J. & E. Hall to fit each cylinder with a safety valve so that no accident can possibly occur.

# CHAPTER V.

## TYPES OF MACHINES.

THE principle of the compression machine has been generally explained in the Introduction. The same cycle is gone through, and the same elements are required both for ammonia and $CO_2$. There are a great many makers of ammonia-compression machines, and each maker has his own patents and peculiarities of construction.

The whole question of refrigerating machinery *per se* has been dealt with by Mr Norman Selfe and other writers, and all that is necessary here is a brief mention of the principal features of the machines made by some of the leading manufacturers.

The question is gone into more fully in the next chapter, but it may be pointed out that the efficiency of a refrigerating machine depends most largely upon the efficiency of the compressor, and that a perfect compressor should be able to discharge on its delivery stroke the whole of the gas taken in on its suction stroke. There should also be no leakage of gas past the piston, and, for reasons subsequently explained, the superheating of the gas during compression should be avoided.

**The De la Vergne Refrigerating Machine.**[1]—The essential difference between the De la Vergne machine and

[1] Messrs L. Sterne & Company, Ltd., Glasgow.

those of other makers lies in the fact that the requirements referred to above are obtained by injecting a measured quantity of mineral lubricating oil into the compressor at each stroke. This oil not only lubricates the compressor, but seals the clearance spaces and takes up the heat developed during compression. A sectional view of the complete De la Vergne plant is shown in fig. 22. An enlarged view of the double-acting compressor is shown in fig. 23. Reference to this will show that all the clearance spaces in the compressor are full of oil, which is injected under pressure on the suction stroke. On the compression stroke, the oil and the ammonia pass away through the delivery valves together. It is essential, in order to avoid the expansion of the compressed gas, that all the ammonia gas should be discharged from the compressor before the oil. On the up-stroke this is simple enough, but to obtain the same results on the down-stroke presented serious difficulties. These have been overcome, however, by an ingenious arrangement of valves shown at the lower end of the compressor.

As soon as the pressure gets sufficient on the down-stroke of the piston, these valves will open to the condenser and the gas will pass away through them. As the piston descends, it will cover first the upper valve and then the lower valve. When the piston is at the bottom of its stroke, however, the upper valve comes into connection with an annular chamber in the piston itself. This chamber has valves in its bottom, and as soon as the lower valve is covered, these open and pass the gas, and then the oil through the annular chamber and the upper valve, and so out of the compressor.

The gas and oil pass away through the delivery pipe to

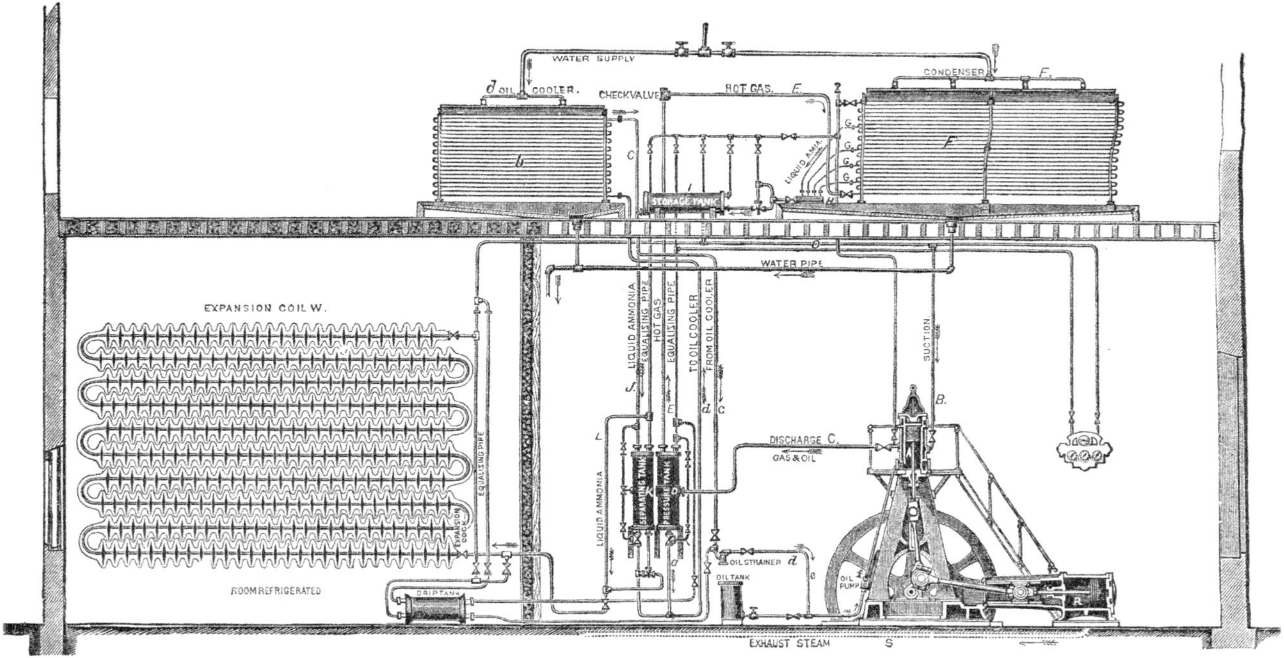


Fig. 22.

To face page 98.

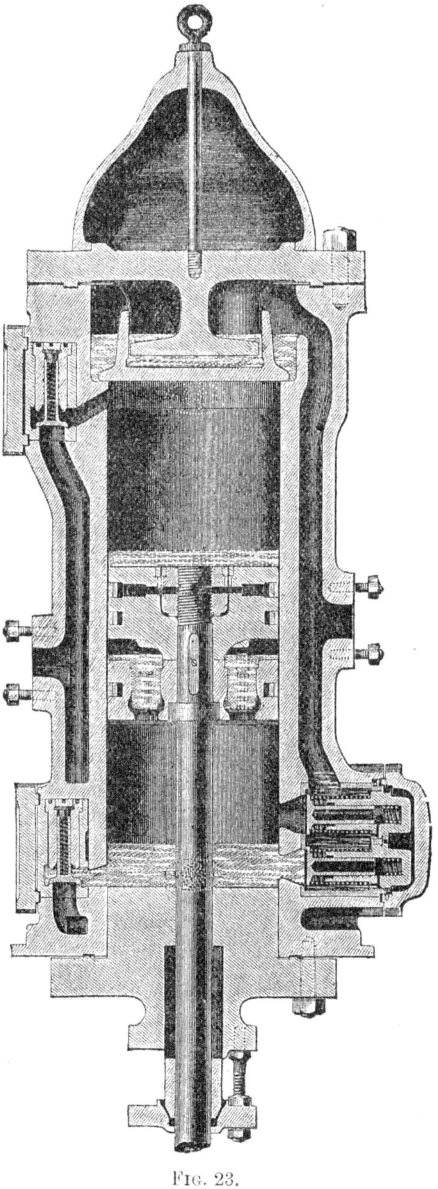

Fig. 23.

the pressure tank; here they separate, the hot gas passes from the top of the tank and goes to the condenser, the oil passes from the bottom and goes to the oil-cooler. The ammonia gas is liquefied in the condenser, and runs off through the small pipes to the storage tank, and from that to the separating tank, where any oil that might have been carried over with the hot gas is finally separated from the liquid. The liquid ammonia passes from this tank to the expansion cock, where it expands, passes through the expansion pipes, where it takes up heat and becomes a gas, and finally passes to the suction valve of the compressor. The oil, after being cooled, flows through the oil-strainer, where any impurities are caught, and is again injected into the compressor.

The illustration shows an oil-pump for this purpose, but in the more modern machine this has been replaced by an oil-injector. The other pipes shown are for equalizing the pressures and for pumping out any portion of the plant, should it be necessary for the purpose of repairs.

**The Frick Company's "Eclipse" Machine.**[1]—The leading characteristic of this machine is that the compressors are single-acting, two of them being arranged side by side and driven by a steam-engine which forms part of the complete machine. The general design of the compressor can be readily grasped from fig. 24.

The under side of the piston is in free communication with the evaporating coils, and the suction valve is placed in the piston itself. The delivery valve is in the compressor head. Owing to the necessity of completely emptying the compressor of gas at each stroke, it is necessary to have an exceedingly fine clearance between the piston and the compressor head. If this clearance is, say, $\frac{1}{32}$ of

[1] Messrs the Frick Company, Waynesboro, Pa., U.S.A.

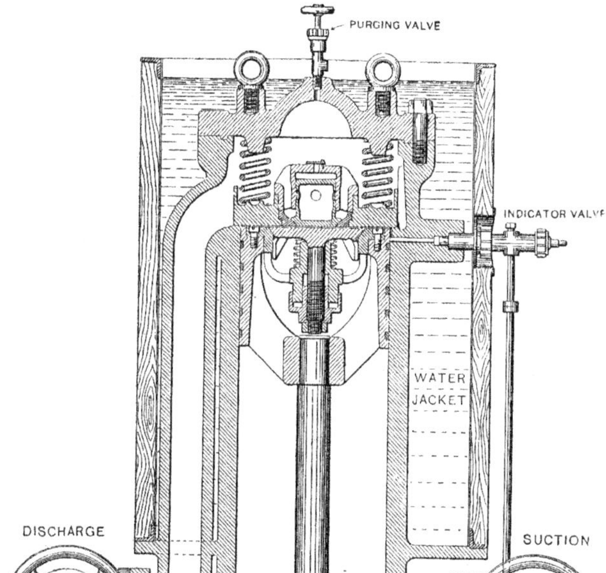

PURGING VALVE
INDICATOR VALVE
WATER JACKET
DISCHARGE
SUCTION
OIL PUMP
PACKING
OIL PIPES


FIG. 24.

an inch, and there is any wear on the connecting rod, a knock might easily be developed, and in extreme cases the compressor head might be knocked out. In order to secure the minimum of clearance, and at the same time guard against danger in this respect, the Frick Company make the compressor head movable, and keep it in place by heavy spiral springs.[1] The head thus becomes a valve of large area, and if a knock does occur, it lifts and no damage is done. This is merely a safety device, and the head is, of course, not intended to act as a valve in normal working. It will be noticed that the compressor head and discharge valve are entirely independent of the compressor casting. In the event of accident this allows the complete head to be quickly removed and replaced by a spare one. The machine is worked on the dry compression principle, of which more subsequently; and the heat of compression is removed by the water flowing through the water-jacket. The compressor gland is of considerable length to prevent leakage of gas. It is kept lubricated by means of a small oil-injector, and the gland can be tightened up by bevel gearing, as shown.

**The Haslam Machine.**[2]—This firm make a large variety of designs, amongst them being a special compound ammonia-compressor for use in hot climates. This is arranged with high and low pressure single-acting compressors arranged either tandem or alongside each other. For use on board ship, the ammonia-compressors are placed with the compound steam cylinders on a cast-iron bed-

[1] In the Enock safety compressor the effect of the safety head is obtained by providing the cross-head with an extended trunk, in which is placed a powerful spring. The piston-rod bears on the top of this spring, and when the piston makes metallic contact at each stroke with the piston cover, the spring is slightly compressed.

[2] Messrs The Haslam Foundry and Engineering Company, Ltd., Derby.

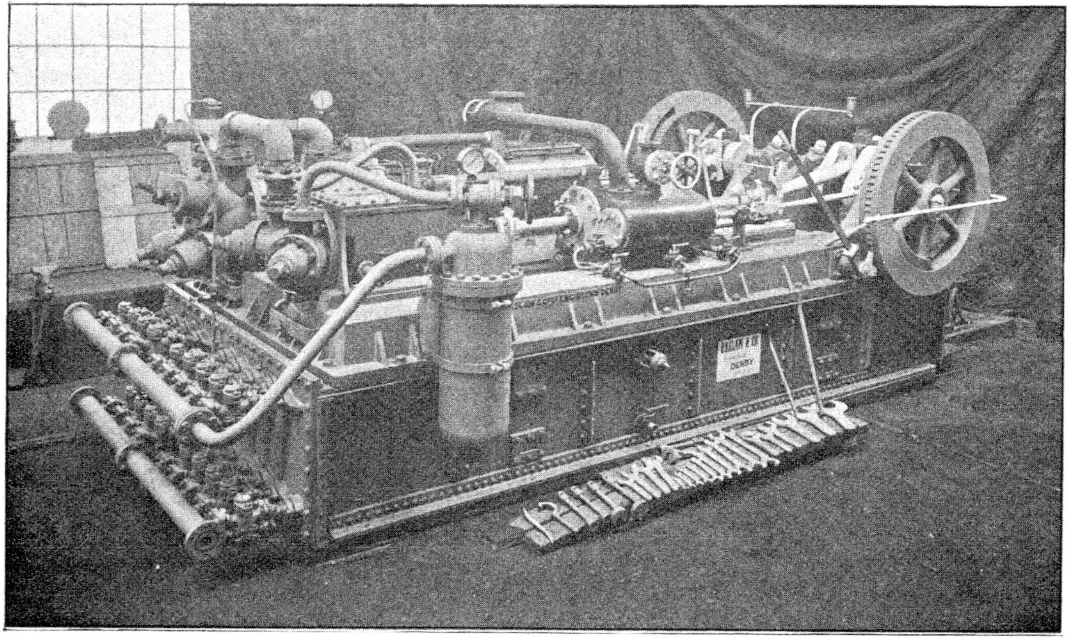


FIG. 25.

*To face page* 103.

plate mounted upon a massive wrought-iron rectangular tank, which contains the ammonia-condenser coils, as shown in fig. 25.

A sectional view of the Haslam ordinary type of compressor for general work is shown in fig. 26. The whole of the cast-iron is of special mixture cold blast with very close grain; the body, as shown, is not water-jacketed, but can be adapted to suit requirements, the end covers are fitted with two or four valves according to size of cylinder, each valve being self-contained with its seating, guide, etc., so that the whole can be withdrawn at one operation and renewed in a very short time without breaking any but its own cover joint; the valves are generally made of the best low-moor iron or mild forged steel with malleable or steel seatings and guides; the suction valve is arranged with a collar on the spindle, so that, in case of breakage, the valve cannot get into the cylinder. The valves are adjusted to their joints by an outside screw, thus obviating the difficulty of making the valve-box cover joint and valve-seat joint with the one set of cover studs. The stuffing-box gland in the front covers is arranged with two separate sets of packings with a lantern between, which can be used either for an oil-seal or connected back to the suction pipe as desired; when the latter arrangement is adopted, the lubrication of the rod is provided for in the small gland at the front, which is fitted with an outer nut and a single turn of soft packing to prevent the oil escaping. The feature of the whole stuffing-box is that either set of main packings can be set down independently. This is necessary, as one packing may be quite tight, while the other may require adjustment. The piston is fitted with four cast-iron rings; the ends of the piston are turned to a template made to suit the end

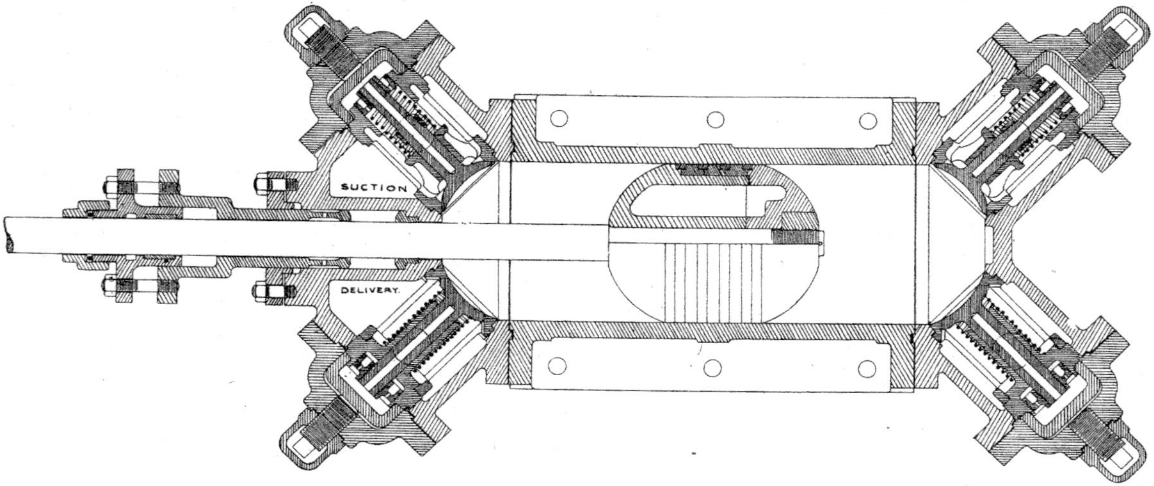

Fig. 26.

covers. By this means the clearances are reduced to a minimum.

**The Hercules Machine.**[1]—A general view of the complete Hercules machine is shown in fig. 27. Of the engine little need be said, except that in the standard types it is horizontal and usually of the Corliss type. The compressors are vertical, but instead of being placed at right angles to the engine and above it, they are arranged alongside, the rotary motion of the engine being changed into a reciprocating motion by the heavy rocking-shaft shown. This rocking-shaft is linked to the compressor piston crosshead, which works in guides formed in the solid casting of the frame. The compressor itself is shown in section in fig. 28. It is single-acting, but has no valve in the piston.

In all compressors, as the suction valve is held up to its seat by a spring, unless it is perfectly balanced there must obviously be pressure behind it to cause it to open; this is common to all valves, and so the pressure of the gas in the compressor is less than the evaporator or back pressure by the pressure necessary to open the valve. In the Hercules machine this is got over in the following way:—The annular chamber at the bottom of the compressor and extending up its side to the suction valve is in direct communication with the suction pipe, and the gas in it is at the full back pressure. When the piston is at the bottom of its stroke, it is also in free communication with the evaporator through the ports in the compressor casting. In the illustration the piston is shown at the bottom of its stroke; as it started to descend, the suction valve opened and the cylinder began to fill with gas in the ordinary way. As it passed the ports, free gas from the annular chamber entered, and the pressure in the com-

[1] Mr C. A. MacDonald, 63 Pit Street, Sydney, N.S.W.

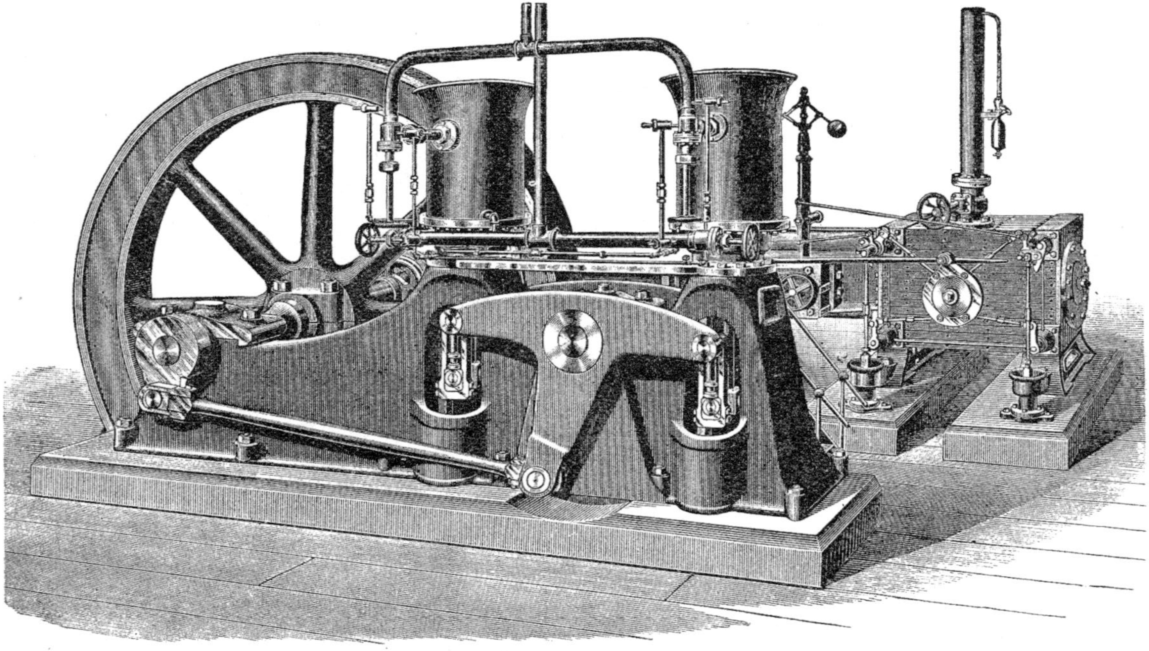

Fig. 27.

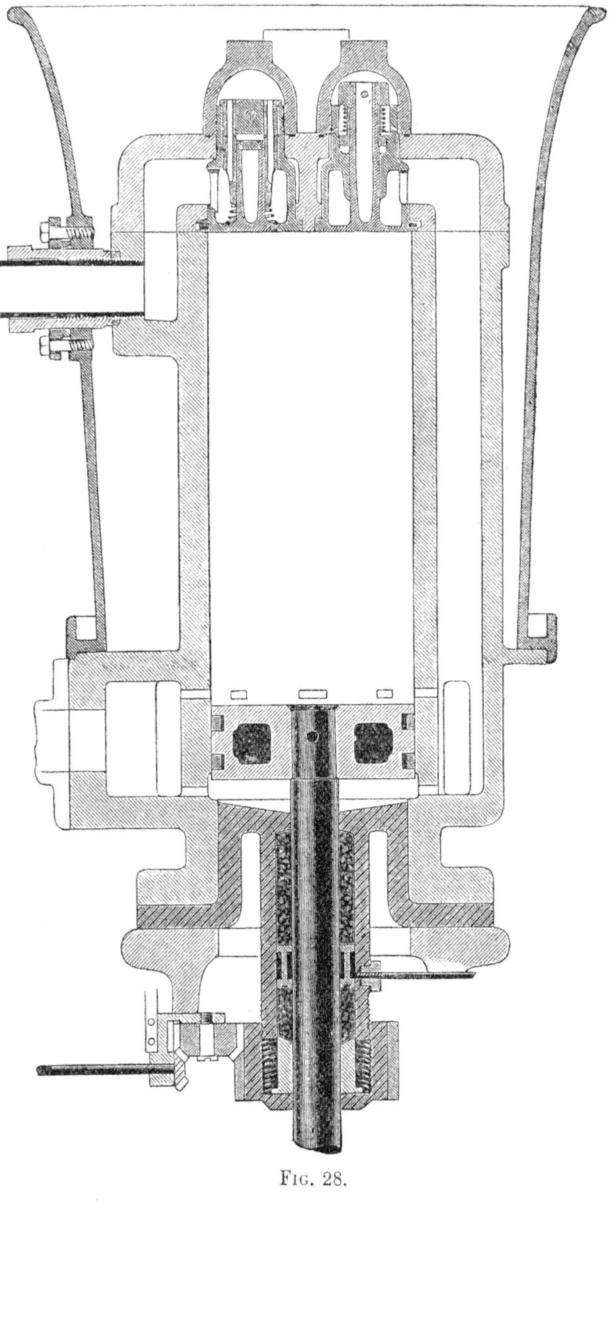

Fig. 28.

pressor was equalized and made the same as that of the evaporator. On its return stroke the piston closed these ports and compressed the gas, discharging it through the delivery valve shown on the right of the illustration. Fig. 28 also shows the stuffing-box with its lubricating and sealing lantern and its geared gland. The funnel-shaped casting on the outside of the compressor is for a water-jacket.

**The Linde Machine.**[1]—This company also makes a large variety of machines each suited for its own purpose. One of the latest type built for use on land is illustrated in fig. 29.

The machine is driven by a compound engine with Paxman drop valves, and is specially built for use with highly superheated steam. The compressor, which is made to work with wet compression, and is therefore not water-jacketed, is shown in section in fig. 30. It is double-acting, the barrel being made of a special mixture of cast-iron of hard close grain, the ends being made concave to allow more room for the valves and to give smaller clearances. The valves are of steel, and each is made self-contained and fitted into a malleable iron box which forms the valve seat as shown. This design makes it possible to withdraw the whole valve for examination or repair by merely breaking the joint of the valve cover and without disturbing any of the connections. The delivery valves are fitted with a special buffer spring, which effectually prevents any tendency to hammer. The stuffing-box and glands are made with an oil-seal and fitted with double packing. The piston is convex to fit the concave ends of the compressor; it is fitted with cast-iron rings, and forms as nearly as possible a metal to metal contact with the

[1] Messrs The Linde British Refrigeration Company, Ltd., London.

Fig. 29.

*To face page* 108.

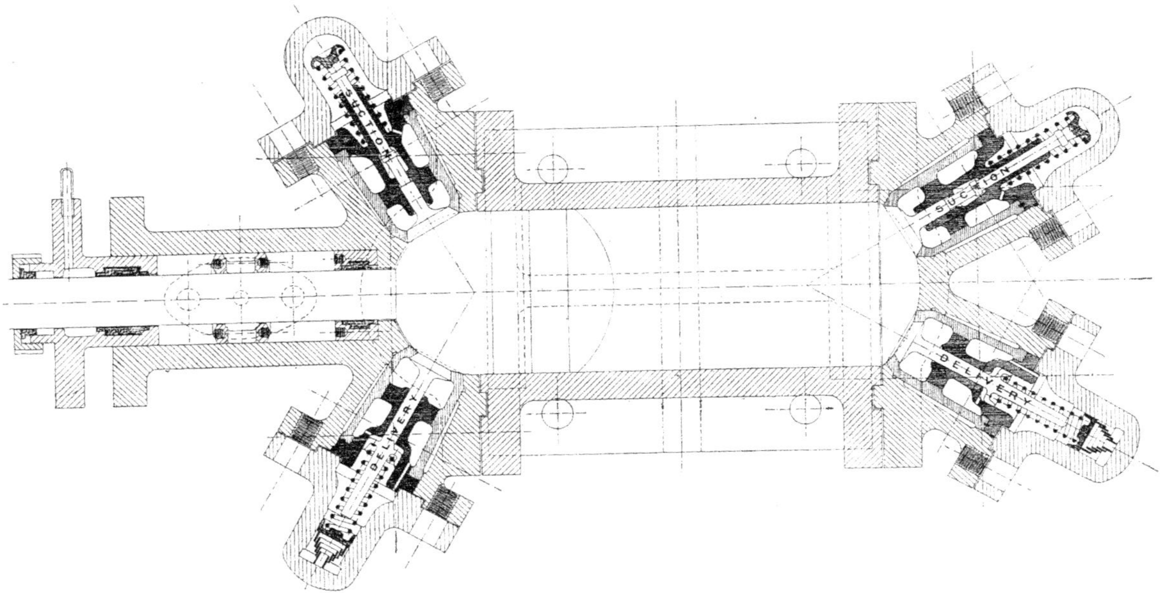


FIG. 30.

cylinder heads. The compressed gas, as it leaves the cylinder, passes through an oil-separator shown in fig. 29, thence to the condenser. Both the illustrations are so largely self-explanatory, that further detailed description is hardly required.

**The Pontifex Compressor.**[1]—The principal feature of this compressor, which is illustrated in fig. 31, is the means taken to prevent the gas superheating during compression. Instead of working with wet compression or of fitting a water-jacket, the cylinder is jacketed by the return ammonia gas on its way from the refrigerator to the suction valves. This enters through the suction valves, passes round the cylinder liner, from which it picks up the heat of compression, and is drawn to the suction valves through the passages shown. The suction and delivery valves are arranged so that they can be completely withdrawn for inspection, etc. without disturbing the connections. The piston-rod gland is lubricated in rather a special way. The oil-separator, in which the oil is of course at the forward pressure, is connected *via* the lantern gland to the suction side by a half-inch pipe, part of which is shown. By regulating the valve on the separator, a continuous stream of oil can be blown from the separator through the gland to the suction ports, where it passes through the delivery valves back to the separator. This arrangement ensures a constant lubrication of the working parts without removing any of the oil from the machine, and has been found to give great economy in the quantity of oil used.

**The Triumph Machine.**[2]—The Triumph machine is of the horizontal type, the engine being placed alongside the

[1] Messrs H. Pontifex & Sons, Ltd., London.

[2] Messrs The Triumph Ice Machine Company, Cincinnati, O., U.S.A.

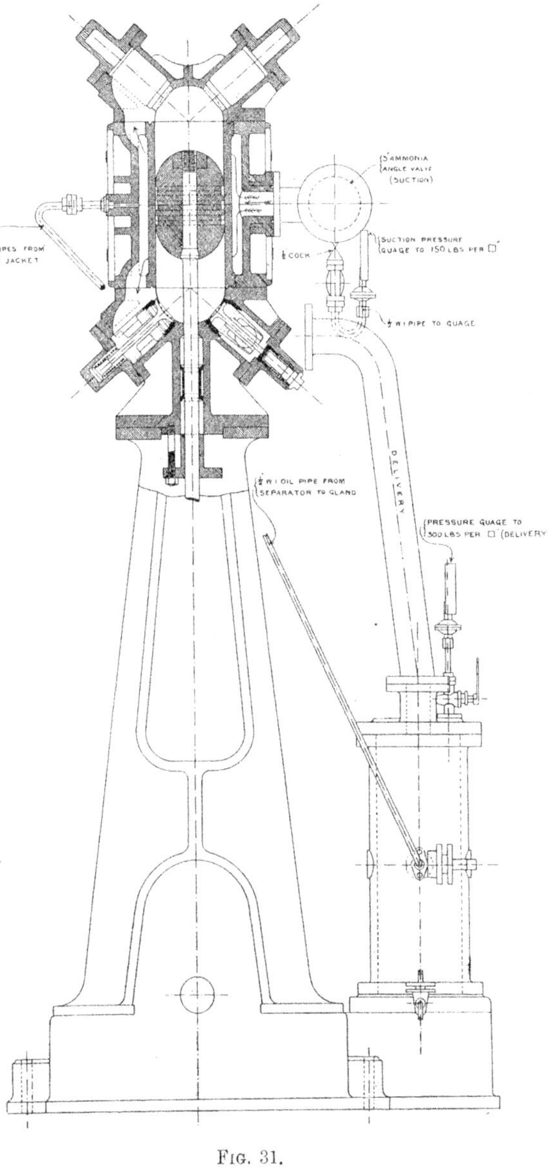

3" AMMONIA ANGLE VALVE (SUCTION)
SUCTION PRESSURE GUAGE TO 150 LBS PER □"
½ COCK
½ W I PIPE TO GUAGE
½ W I OIL PIPES FROM GLAND TO JACKET
½ W I OIL PIPE FROM SEPARATOR TO GLAND
DELIVERY
PRESSURE GUAGE TO 300 LBS PER □" (DELIVERY


FIG. 31.

machine parallel to the compressor. The motion is transmitted through the crank shaft on which the fly-wheel is mounted. The engine is, or can be, of any desired type, with corliss or other automatic valves for large sizes and slide valves for the smaller sizes. Several patterns of machine are made, including small machines driven by electric motors. A section of the compressor is shown in fig. 32. It has concave ends, the piston being made convex to fit them.

There are five valves—three suction and two delivery. The third suction shown at the side of the compressor is lighter than the other two. It is perfectly balanced, and is designed to equalize the difference between the suction pressure and the pressure in the compressor caused by the heavier springs on the two end suction valves. The compressor is surrounded by a water-jacket cast in the compressor-casting itself. The construction of the long lantern gland is clearly shown in the illustration.

### Carbonic Anhydride Refrigerating Machines.

The $CO_2$ refrigerating machine is of comparatively recent date. A machine working with this agent was patented in 1881 by Raydt, and, curiously enough, Professor Linde, whose name is so well known in connection with ammonia machines, was one of the first, if not the first, to experiment with $CO_2$ as a refrigerating agent. One of the first really practical machines, however, was invented in 1886 by Windhausen of Berlin, whose labours in other fields of mechanical refrigeration have already been noticed; and as recently as 1889 J. & E. Hall, who have made the subject specially their own, built their first machine under his patents and to his designs. Since then they have made considerable improvements on these designs, many patents

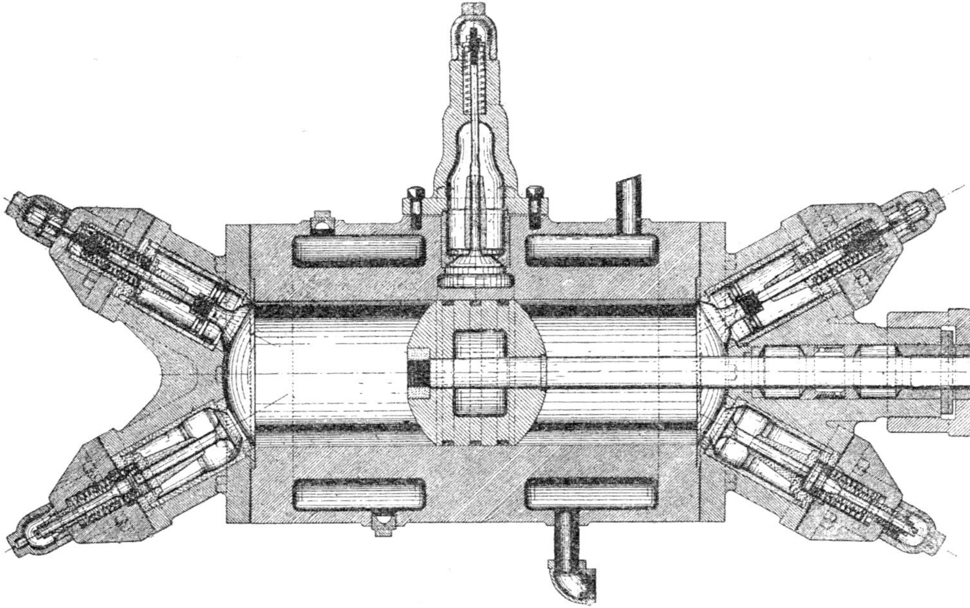

Fig. 32.

have been taken out, and owing to the way in which they have developed the system the $CO_2$ machine now stands high as a refrigerating appliance. The essential parts of the machine consist, like the ammonia-compression machine, of a compressor, a condenser, an expansion valve, and an evaporator. There is, however, a wide difference in the details. The relative behaviour of ammonia and $CO_2$ machines under different conditions has been fully gone into in Chapter II., and the results there obtained represent the theoretical advantages and disadvantages of the two agents.

The practical advantages and disadvantages are quite another matter. Reference to page 90 will show that while the pressure in lbs. per square inch of $CO_2$ at zero F. is 310, that of ammonia is 30, but that for an equal refrigerating effect the relative volumes of the compressor necessary for the two substances is respectively 3·24 and 23·3. Thus, while ammonia vaporizes at a much lower pressure than $CO_2$, it requires a very much larger compressor area. These two points principally govern the design. The high pressure required is merely a mechanical effect, and can be easily and readily dealt with by mechanical means. Any disadvantages it may present are amply compensated for by the smallness of the necessary compressor area and the consequent saving in material and space.

The success of the system has recently caused other makers to place $CO_2$ machines on the market, but as the variation in design is not nearly so great as that to be found in ammonia machines, the description of the type manufactured by Messrs Hall[1] will be sufficient.

The practice with these makers is for the small machines

[1] Messrs J. & E. Hall, Ltd., Dartford.

FIG. 34.

To face page 115.

to be built with vertical compressors as shown in fig. 33 and to be arranged for either belt- or steam-driving; and for the larger sized machines to be horizontal as shown, with simple, compound, or triple expansion steam cylinders

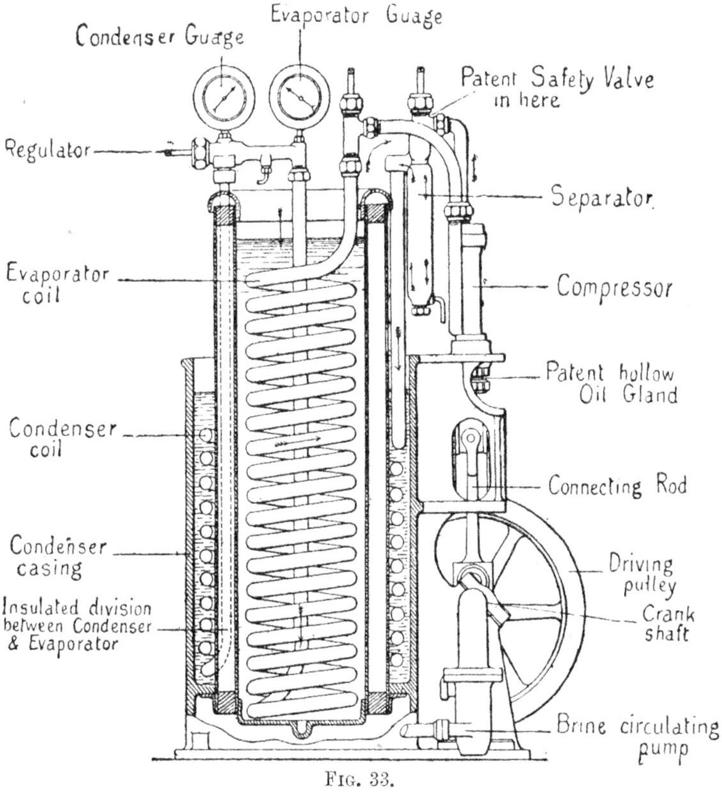


FIG. 33.

arranged tandem with the compressors, which are driven by tail-rods from them. The general design will be readily gathered from fig. 34 which illustrates a 120-ton refrigerating $CO_2$ machine.

The steam and $CO_2$ condensers are sometimes arranged in the bed-plate and sometimes separately, and the compressors, when in duplicate, are so arranged that they can either be run together on the same work, or can be put on to separate duties, so that each compressor and steam cylinder to all intents and purposes comprises a separate machine. This feature is useful in guarding against breakdown, and affords an additional safeguard if the whole machinery is not installed in duplicate.

**The Compressor.**—Reference to the curve giving the temperatures and corresponding pressures of $CO_2$ on page 90 will show the high pressures to be dealt with. In consequence, great soundness and strength of the various parts is essential. To meet this, the compressors are, for the larger machines, bored out of a solid steel forging, cast-iron or cast-steel being unsuitable on account of its porosity. This precaution is taken for the double purpose of securing a sound and reliable material, and of providing a perfectly smooth bore, as it is plain that any unevenness or inequalities in the cylinder would, with the pressures to be dealt with, be a very serious matter. For small machines the compressors, fig. 35, are of bronze ($CO_2$ has no deleterious action on copper). The suction and delivery valves are all made to the same design and are interchangeable. The compressor piston is packed with cup leathers similar to the ram of a hydraulic press, which makes the most gas-tight packing possible. The lantern gland is also made gas-tight by cup leathers as shown, and sealing-oil, at a higher pressure than the highest in the compressor, is forced into it between these cup leathers, keeping them tight against the piston-rod. The superior pressure is maintained on the oil by means of a patent differential piston, having one side in contact with $CO_2$ at the condenser pressure, and the other in contact with the oil which has been forced in against it by a hand-pump. As the area of the piston in contact with $CO_2$ is greater than that in contact with the oil by the area of the piston-rod, a greater pressure per square inch is maintained on the latter than on the former. The oil is connected to the compressor gland, and as any oil leaks past the cup leathers either into the compressor or to the atmosphere, it is at once replaced by oil from this cylinder. When the cylinder becomes empty, oil is again forced into it by the small hand-

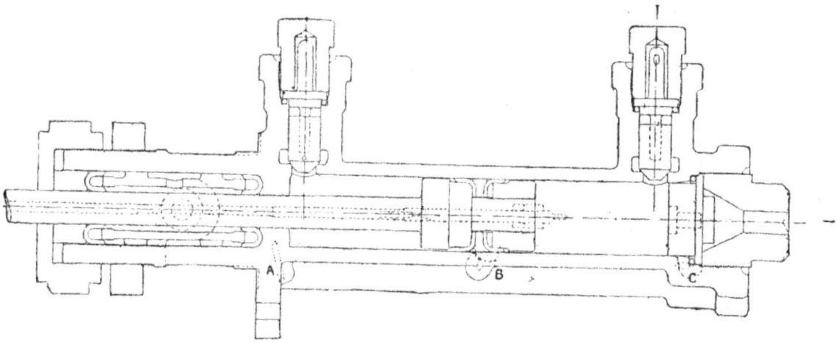


FIG. 35.

pump. A small leakage of oil into the compressor is an advantage, as it lubricates the piston and fills up the clearance spaces; in ordinary working, however, the amount of leakage in this way is small, and only a few strokes of the hand-pump are required every four or five hours to replenish the reservoir. The gas discharged from the delivery valves carries with it a certain amount of oil, which is separated before the gas passes to the condenser.

The compressor is fitted with a patent safety valve formed of a copper disc, which is designed to give way at a pressure of some 1950 lbs. on the square inch. In conjunction with this disc, fig. 36 is an ordinary spring safety valve, which closes as soon as the excessive pressure has been relieved. This safety valve is, of course, provided to guard against accident due to a careless driver starting up the machine without opening the valves on the delivery side. In the case of rupture, the $CO_2$, being cheap and harmless, is allowed to escape to the atmosphere. Safety valves have been fitted on some ammonia machines, but in that case, the gas, being more valuable and more pungent, is blown off and passed into the suction side of the plant.

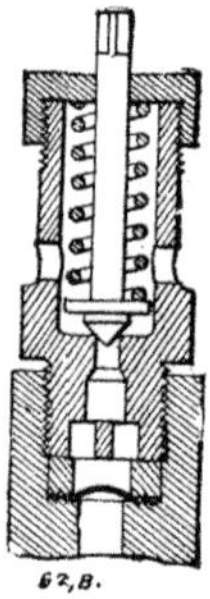

FIG. 36.

**The Condenser.**—The condenser is made of both the submerged and atmospheric type. Owing to the pressure, special hydraulic piping, usually of 13/16 bore, is employed, the lengths being electrically welded together, so as to minimize the number of joints generally, and to render it unnecessary to have any joints in an inaccessible position.

These condenser coils for land work are usually of wrought-iron, but, for marine work or for any special purpose, the coils, both in the condenser and evaporator, can be made of copper either plain or tinned.

**The Evaporator.**—The evaporator is, in general design, similar to the condenser. For cold storage work and ice-making, separate brine-coolers are almost invariably employed. In this case the evaporating coils are nested one inside the other. In cold stores employing the air-blast system the air is blown directly over brine coils in much the same way as that described in Chapter XII., or in some cases over larger metal surfaces wetted and consequently cooled by the cold brine from the evaporator.

**General Considerations.**—It has frequently been urged

that $CO_2$ machines cannot work successfully under conditions found in tropical and semi-tropical countries, where the temperature of the condensing water is high. The incorrectness of this contention will be apparent from a careful examination of the entropy diagrams in Chapter II., and in this case theory is fully borne out by the results which have been obtained in practice. The compactness of the $CO_2$ machine makes it especially suitable for work in positions where space is of especial value, such, for instance, as on board ship. The smallness of the charge and the working parts, and also the existence of the safety valve, renders a serious explosion practically impossible. The harmlessness of the gas prevents any escape that may occur from having serious consequences.

# CHAPTER VI.

## THE REFRIGERATING PLANT.

**The Compressor.**—Experience with compressors has now been so considerable that there is very little to choose between several designs, and the only room for variation is in the piston and valves. It is very important that these should be as perfect as possible, and it is well to have the area of the suction valves rather greater than that of the delivery. The piston speed should be low—about 200 feet per minute—and the clearance should be reduced to vanishing point. The glands in common use by various makers are illustrated in Chapter V.; they should be sufficiently long to ensure that the piston-rod exposed to the gas inside the piston never leaves the gland and is not exposed to the external atmosphere. Needless to say, the compressor should be firmly secured to its bed with turned bolts and steady pins. If there is any vibration or movement, it will be impossible to work to fine clearances.

**Testing for Clearance.**—A very simple way of testing the clearance of a compressor piston is by taking out one of the valves back and front, and inserting a piece of sheet-lead against the cylinder head. By barring the engine round, the piston is brought against the lead, and squeezes it out to the thickness of the clearance.

If it is undesirable or inconvenient to remove the valves, the engine can be barred round until the compressor crank is on the back centre. Mark the position of the cross-head carefully, then uncouple the connecting rod and push the piston as far back into the cylinder as it will go, and measure the difference. This will represent the clearance. Repeat the process with the front end. Indicator diagrams, of course, give an idea, but it is not always possible to tell by them how much is clearance.

**Superheating during Compression.**—In compressing gas in a cylinder, work is done, and a certain proportion of the work converted into heat, which reveals itself by raising the temperature of the gas during compression, and by raising the temperature of the cast-iron cylinder, its valves, piston, and piston-rod, and all parts in contact with the hot compressed gas. On again filling the cylinder through the suction valve, cold gas of a certain temperature and volume is drawn into the hot compressor, and will immediately absorb some of the heat from the cylinder walls, piston, and rods, thus raising its temperature, and increasing its volume. The greater the temperature of the gas, the smaller the weight that can be contained in a given capacity; consequently the heating of the gas during admission to the compressor results in a less weight of gas being compressed per stroke. As the efficiency of the compressor is measured by the weight of gas that can be passed through it per stroke, it is evident that the cooler the parts coming in contact with the incoming gas can be kept, the lower the resulting temperature of gas before compression, and, therefore, the greater the weight for a given volume.

The best means of obtaining this result have been the subject of a wordy warfare, which has raged around the subject for many years. It is not proposed to offer an opinion, but merely to point out the four principal methods which have been adopted for combating the evil. These are:—

1. Double-acting compressors, with wet compression and freezing back.
2. Do. do., with oil-injection and dry compression.
3. Do. do., with water-jacket and dry compression.
4. Single-acting compressors, with or without 2 and 3, and with dry compression.

**Wet Compression.**—With wet compression a certain amount of liquid ammonia is injected into the compressor with the gas, or, expressed more correctly, the gas is returned to the compressor in such a highly saturated state that it carries over a certain amount of liquid with it. As the object is to cool the compressor down and prevent superheating, this liquid is, or should be, just sufficient, by evaporating in the compressor, to absorb the heat of compression and allow the compressed gas to leave the compressor at a temperature of somewhere about 98° F., or just below the point of saturation. It is easily seen that with wet compression it is absolutely imperative that no more liquid should be introduced than will just suffice, by evaporating, to cool the compressor and piston to the required temperature. There is always a loss of capacity in an ammonia-compressor, due to the gas which remains in the clearance spaces expanding on the return stroke of the piston, and destroying what, for the sake of brevity, may be called the vacuum; and if, instead of this gas, the clearance space contains liquid, which will on the return stroke of the piston expand into gas and fill the compressor, the loss of capacity will obviously be much greater. Freezing back is much more necessary with double-acting than with single-acting compressors, because the piston and cylinder in the former case get much hotter than in the latter, and the gas, as it enters through the suction valves, is expanded to a correspondingly greater degree.

**Oil-Injection.**—Instead of freezing back to any great extent, the De la Vergne Company introduce a measured quantity of oil into the compressor at each stroke. This oil, which has already been referred to in Chapter V., serves the triple purpose of partially absorbing the heat of compression, sealing the clearance spaces, and lubricating the cylinder. After being heated in the compressor, it is separated from the ammonia, cooled, and again injected. When either this or the two following methods are in use, no great quantity of liquid is, if it can be avoided, introduced into the compressor, and the engine-driver strives to return his gas in a nicely saturated condition, so that the suction pipes are—or would be if they were not insulated—lightly frosted over right up to the suction valves. The compressed gas, as it leaves the delivery valves of a dry compressor, should heat the pipes to a temperature which is not uncomfortable when the back of the hand is laid upon them.

**Water-Jackets.**—Except in machines where oil-injection or wet compression is in use, water-jackets are usually provided; their object, of course, being, as stated above, the removal of the heat of compression. The use of a water-jacket is an economical method of attaining the object in view, and most makers recognize this by adopting it.

**Single-acting Compressors.**—So far as the avoidance of superheating is concerned, single-acting compressors are recommended by some authorities on the ground that they do not get so hot, and, therefore, do not cause the same loss of capacity. They are, however, frequently water-jacketed. The first designs of the De la Vergne compressor with oil-injection were exclusively for single-acting compressors; this, however, was due, not so much to an acknowledgment of their superiority, as to the mechanical difficulties, which had to be, and now have been, overcome, in introducing a double-acting compressor.

**Indicating the Compressor.**—The method of indicating a compressor, the indicator used, the drum motion, the reducing lever, etc., is similar to those in every day use for a steam-engine, the only difference being that, for ammonia, the indicator, etc. must not be of brass as is common with steam-indicators.[1] At first sight a diagram from an ammonia-compressor is very much like that from a steam-engine. The essential difference, however, is that while in a steam-engine the steam is expanding and doing work, in the compressor the ammonia gas is having work done upon it. Thus, in following a steam-engine diagram, the eye travels round the diagram with the clock; in the compressor diagram it travels against the clock. Fig. 37 illustrates a good diagram from an ammonia-compressor.

The fact that the valves are working well and leave ample area is evidenced by the gauge pressure corresponding nearly with the pressure of delivery at A A, and by the suction line agreeing with the refrigerator pressure of ½ lb. The clearance is very small, only a slight heel showing at the right front end B. The low back pressure and

[1] Good ammonia-indicators are made by nearly all the indicator-makers, such as The Crosby Steam Gauge & Valve Company, Messrs Schaffer and Budenberg, etc.

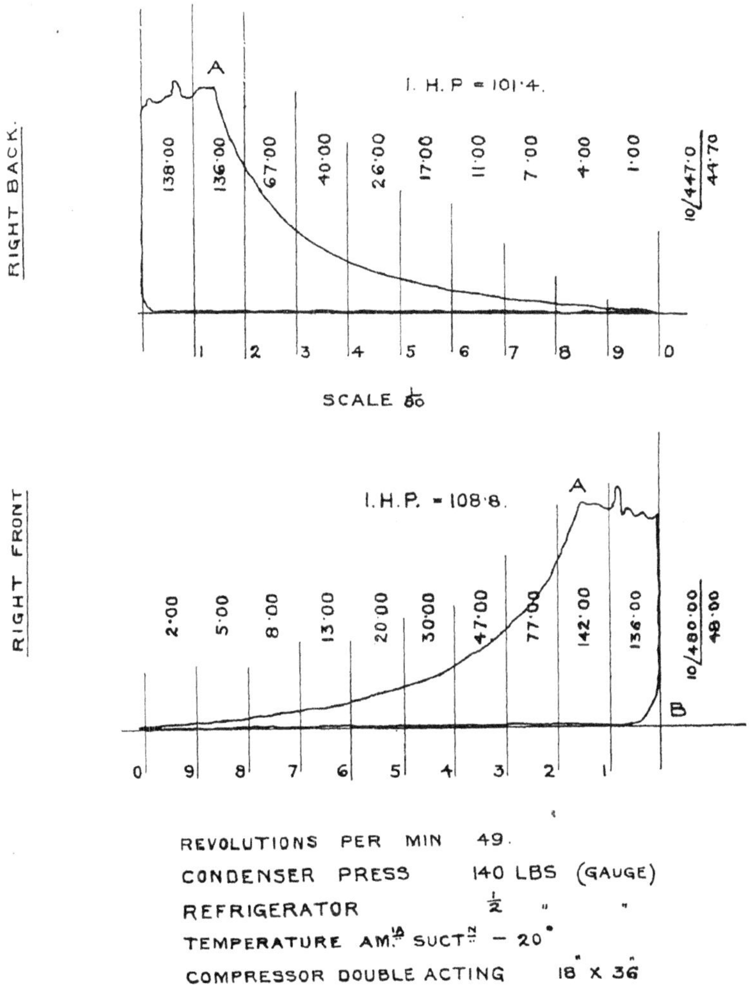
RIGHT BACK.
A
I. H. P = 101·4.
138·00
136·00
67·00
40·00
26·00
17·00
11·00
7·00
4·00
1·00
10/447·0
44·70
SCALE 1/80
RIGHT FRONT
A
I.H.P. = 108·8.
2·00
5·00
8·00
13·00
20·00
30·00
47·00
77·00
142·00
136·00
10/480·00
48·00
B
REVOLUTIONS PER MIN 49.
CONDENSER PRESS 140 LBS (GAUGE)
REFRIGERATOR ½ " "
TEMPERATURE AM.IA SUCT.N − 20°
COMPRESSOR DOUBLE ACTING 18" X 36"

Fig. 37.

temperature is accounted for by the fact that the machine from which this diagram was taken is employed in maintaining a very low temperature.

Fig. 38 is another diagram taken from a machine cooling

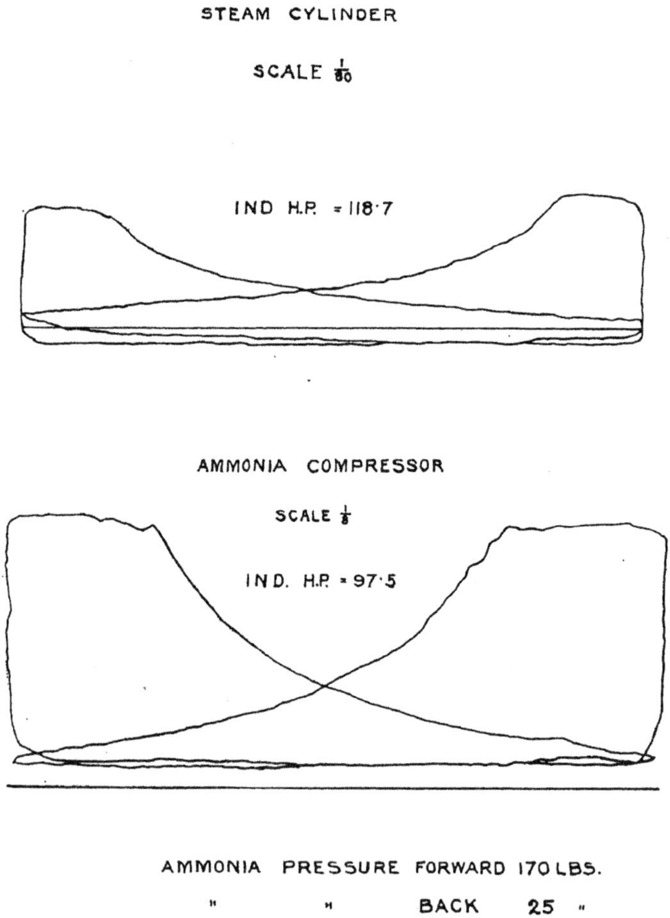


FIG. 38.

liquor, which accounts for the high back pressure. The diagrams are fairly good, but the fact that the compression line does not project above the pressure line, but is slightly below it, rather points to a leak in the delivery valves. The piston would also appear to be leaking slightly, as evidenced by the shape of the heel of the cards. The engine was also used for driving a fan and pump, which accounts for the big difference in the I.H.P. of the steam-cylinder and compressors. The usual excess is from 12 to 20 per cent.

Fig. 39 illustrates a very bad diagram, the disease

SCALE $\frac{1}{60}$

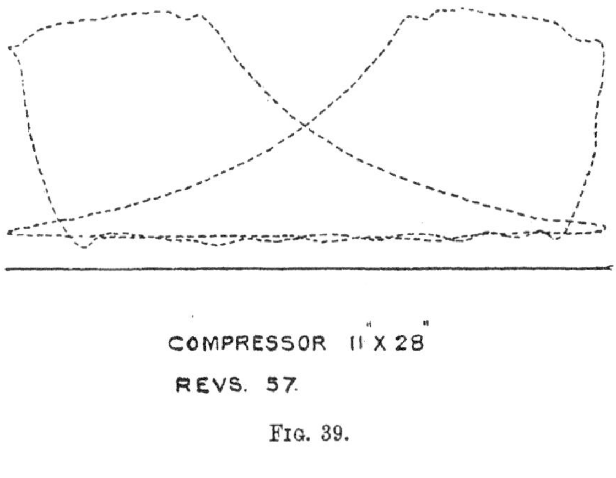

COMPRESSOR 11" X 28"

REVS. 57

FIG. 39.

being excessive clearance, which is indicated by the flat course of the back line. The compression line is not good either, and the springs on the delivery valves are evidently much too strong. This condition of affairs is indicated by the pressure falling towards the end of the stroke when the piston speed is low, showing that an increased pressure is demanded at the highest piston speed on account of the valves being too small. Familiarity with indicator diagrams soon renders it easy to diagnose the complaints of the compressor, and points at once to the cause of the trouble or to detect faults that might otherwise exist undiscovered.

**Horse-power, etc.**—Though the foregoing are the principal objects of indicating a compressor, it is also useful at times to ascertain the relative horse-power of the engine and compressor. This can be found from the card by the ordinary method of dividing the card up into vertical divisions and obtaining the mean effective pressure (M.E.P.) in lbs. per square inch by measuring off the mean pressure in each of the divisions with a scale corresponding to the strength of the spring used in the indicator, adding them together, and dividing by the number of divisions, as in fig. 37. When the M.E.P. has been found, it will be necessary to multiply it by the net area[1] of the piston, the mean speed of the piston in feet per minute, and divide the product by 33,000, the number of foot-lbs. in a horse-power, thus:—

$$\frac{\text{M.E.P.} \times \text{piston speed} \times \text{piston area}}{33{,}000} = \text{I.H.P. of compressor.}$$

If there is no scale handy, the M.E.P. can be easily ascertained by marking off the height of the vertical division continuously on a piece of paper, measuring the total length of these divisions in inches by an ordinary 24-inch rule, dividing by the number of divisions as before, and multiplying the quotient by the scale of the spring. The simplest way of all is to ascertain the area of the

[1] This is usually the full area behind and the area less the area of the piston-rod in front. If the compressor or engine has a tail-rod, the area of the piston-rod must be deducted in both cases.

card in square inches by one of Amsler's planimeters (they can be obtained from any good instrument-maker, and No. 3 is the most suitable), divide the area by the length of the diagram in inches and decimals, multiply this by the scale of the spring, and the result will be the M.E.P. of the diagram.

There are any number of tips for economizing labour in working out diagrams, forming constants, etc., but it would take too much space to set them out here.

**Ice-making Capacity of Compressor.**—The ice-making capacity of any given compressor varies with the pressure of the gas going into the cylinder. The higher the back pressure, the heavier the gas, and consequently a greater weight is passed through in a given time. To illustrate this:—

Assume that ice is being made with the gas entering the compressor at a gauge pressure of 20 lbs., 15 lbs., and 10 lbs. respectively, the average temperature of condensing water being 70° F., and the compressor 12 inches diameter by 24 inches stroke double-acting running at 60 revolutions per minute,

$$\left.\begin{array}{r}\text{The capacity of compressor}\\ \text{per minute in cub. ft.}\end{array}\right\}\frac{113 \times 24 \times 60 \times 2}{1728} = \overset{\text{cub. ft.}}{188.}$$

Thus it is shown in the table on p. 130 that the ice output of the compressor at 10 lbs. back pressure is barely 72 per cent. of its output at 20 lbs. back pressure.

The calculations in the table are theoretical, and from 20 to 25 per cent. must be allowed for losses in the compressor due to superheating and clearance, and 5 to 10 per cent. for other losses. A compressor capacity of 16,000 cubic inches per minute per ton ice made per 24 hours is a ready rule for obtaining the ice-making capacity of any compressor in England, averaging the yearly output.

| | Gauge Pressures. | | |
|---|---|---|---|
| | 20 lbs. | 15 lbs. | 10 lbs. |
| Volume of ammonia in cub. ft. per lb. of weight, . . | 7·89 | 9·02 | 10·86 |
| Refrig. effect of 1 lb. of ammonia latent heat − (average T. of condg. water − refrig. temperature × specific heat), . . . For ex.: $551-(70-6\times 1)=487$. | B.T.U. 487 | B.T.U. 485 | B.T.U. 482 |
| Lbs. of ammonia circulated per min. $\frac{\text{cub. ft. capacity}}{\text{vol. per lb.}}$, . | 23·82 | 20·8 | 17·3 |
| B.T.U. extracted per min. lbs. × B.T.U., . . . | 11,600 | 10,088 | 8338 |
| Theoretical ice-making in tons per 24 hours $\frac{\text{B.T.U. per min.}\times 60\times 24}{600{,}000}$, | 27·8 | 24·2 | 20·0 |

**Ice-making Capacity and Refrigerating Capacity.**—The ice-making capacity of a plant means the number of tons of ice that can be made per 24 hours under average conditions—that is to say, water at a normal temperature of 60° cooled down and frozen into ice at a temperature of, say, 20° F.

In this case the heat units to be extracted from the water at 60° F. would be $60-32=28$. Add to this the latent heat of ice, or 142·5; the heat to be removed = 170·5 B.T.U.

The latent heat of ice is ·5, so that 32 − 20 × ·5, or 6 B.T.U. have still to be added, making a total of 176·5 B.T.U.

The refrigerating capacity of a machine is the capacity for absorbing heat equivalent to that possessed by one ton of ice at 32° F. melting into water at 32° F. With British practice this is equivalent to an absorption of 318,080 B.T.U. With American practice it is only equivalent to 284,000 B.T.U. The difference is due to the fact that an English or long ton is 2240 lbs., and an American or short ton 2000 lbs.

In practice, owing to the losses mentioned above, the actual ice-making capacity is only about one-half of the refrigerating capacity.

## THE CONDENSER.

The function of the condenser is to get rid of the heat of liquefaction and compression, and thus to liquefy the gas passing through it. Condensers are merely devices for exchanging heat from one substance to another, and it must be borne in mind that the greatest theoretical efficiency can be obtained when the fluid or gas to be condensed comes into direct contact with the condensing medium. This, however, is impossible of adoption in those cases where it is necessary to prevent the fluid or gas from intermixing with the condensing medium, and, therefore, it becomes necessary to introduce separating walls between the two; so that the efficiency of the condenser will now depend—all other things being equal—on the capacity for the transmission of heat possessed by the material of which these separating walls are composed. This, however, is not all.

It has been found that[1] the transmission of heat is practically independent of pressure; that for a given range of temperature the heat transmitted is nearly proportional to the velocity of the cooling water; and that with the same temperature, range, and velocity more heat is transmitted at the higher temperature. The moral of this is plain—make the condenser of as good a conductor as possible. Pass the water over the condenser surfaces rapidly, and let the warmest water meet the warmest gas, and the coldest water the coldest gas.

On the efficiency of the condenser of an ammonia or $CO_2$ refrigerating machine depends in a great measure the economical working of the plant. Practical considerations, such as, first, cost and space, must have an important influence on design, but, where other considerations are not outweighed by these two, it is good practice to have as large a condensing surface as is within reason possible. A glance at the curve of ammonia, fig. 20, will show the necessity, with this agent, of keeping the condensing temperature, and so pressure, as low as possible. The same thing applies to $CO_2$, but there the rise in pressure is less rapid.

**Types of Condensers.**—Refrigerating condensers are of three kinds:—

(1) Submerged.
(2) Atmospheric or evaporative.
(3) A cross between the two, or a double-pipe condenser.

The submerged condenser, as its name implies, consists

[1] Stanton on "The Passage of Heat between Metal Surfaces and Liquids in Contact with them," *Roy. Soc. Phil. Trans.*, 1897.

of a coil or coils of pipe, connected by manifolds at the inlet and outlet and placed in a tank. The hot gas enters the condenser at the top, and the liquid runs away at the bottom. The condensing water follows the opposite course, enters cold at the bottom, and flows out warm at the top.

The atmospheric condenser consists of a coil of pipes containing the medium, and over which water is kept trickling. They are preferably mounted on the roof of the building, or in some other open position, in order that the air, by evaporating some of the condensing water, may assist the cooling and liquefaction of the gas.

**The Submerged Condenser.**—In the submerged condenser the coldest water meets the coldest gas, and the temperature range is maintained, but the movement of the water over the coils is sluggish, and does not assist the rapid interchange of heat, and for this reason more water and more surface are required than with the other type.

The nesting of the coils and the contracted space make cleaning difficult, and small bubbles of air are apt to collect on the coils, which by their insulating effect further retard the heat exchange.

For marine work the submerged condenser, owing to its enclosed form, is the most suitable type. This, and the small amount of space occupied, seem to be the chief advantages it can claim, though, undoubtedly, where compactness is desired, as in a self-contained plant, they are considerations of great importance. One great disadvantage of the submerged condenser is that, owing to the readiness with which water absorbs ammonia, a leak, unless serious, is never discovered, as few engineers take the trouble to test their condensing water. As an illustration of this, on one occasion a coil in a submerged condenser on shipboard split, the ammonia began to escape into the sea-water, and the driver was unable to account for the deficiency in the plant. The whole charge was only prevented from escaping by the formation of sal-ammoniac, which stopped the leak. Another point is that in order to clean or repair a submerged condenser, the coils have to be taken out of the tank, which usually involves great trouble.

**The Atmospheric Condenser.**—This type is by far the most common, and there are nearly as many designs as there are makes of refrigerating machinery. The most important types have been described and illustrated elsewhere.[1] The object in nearly all is to so arrange matters that the cold water, as it runs on to the condenser, comes in contact first with the liquid ammonia and then with the hot gas. This is admirably carried out in the De la Vergne condenser shown in fig. 22, page 98. The hot gas is taken through the pipe *E* (which is carried up above the level of the condenser in order to avoid the liquid running back), and enters the condenser at the bottom. Passing upwards through the 2-inch coils, it is condensed by the down-flowing water, and, as it liquefies, runs away through the small pipes *G* into the header *H*. From this it runs to the separating tank placed on a lower level. The condensing water is pumped into the pipe *F*, and descends in a thin stream over the surface of the coils. Feathers of hoop-iron hang between the coils and guide the water in its descent. These have the disadvantage that to a certain extent they prevent the air passing between and over the coils, and so hinder the evaporative action of the atmosphere. In some instances the feathers have been perforated with ¼- or ⅜-inch holes, and this appears

[1] See Selfe's *Machinery for Refrigeration.* Rich, Chicago, Ill., U.S.A.

to be an advantage. Other atmospheric condensers are made of 1¼-inch pipe continuous throughout, laid in lengths and coiled round at the ends instead of being jointed; in these the gas enters at the top and the liquid runs off at the bottom.

The evaporative condenser has been used to some slight extent where water is scarce, but up to the present it has not come into general use. It is very much the same as the condenser just described, but, instead of trusting to nature for the wind, an air current is created by a fan, and the condensing water is used over and over again. It is stated that good results have been obtained in this way with a small surface.

**The Double-Pipe Condenser.**[1]—This condenser has been recently introduced, with the object of avoiding the defects of both the atmospheric and submerged condenser and of consolidating their advantages. It consists of a 1¼-inch pipe encased in a 2-inch. The gas enters the annular space between these two at the centre of the top coil and spreads both ways until it reaches the centre of the second coil, where it enters the third, and so on. The condensing water is inside the 1¼-inch pipe, which is continuous throughout. It enters the condenser at the bottom and escapes at the top. The flow of water is thus in an opposite direction to the flow of gas, and the coldest water meets the coldest gas. No water passes over the outside of the coils, so that there is no splashing, and the condensers can be placed in any desired position.

**Getting Air out of Condensers.**—It is of great importance that none of the available condensing space should be occupied by air or other permanent gases, and provision has to be made for extracting them from the plant. For

[1] Messrs Westerlin & Campbell, Chicago, Ill., U.S.A.

this purpose a purge-cock is provided at the top of the condenser, with a small pipe leading down below the surface of water in the tray. When extracting the gases, the liquid-cocks of the condenser should be closed and the machine kept running, with water flowing over the condenser. When the pressure has got fairly high, say 200 lbs., the machine should be stopped and the hot gas-cock to the condenser closed. The water should still be kept running. The pressure will not fall much, but as soon as the gauge has come to rest, open the hot gas-cocks and start running very slowly; the pressure should then mount fairly rapidly. When this occurs, stop the machine and close the hot gas-cock. If the purge-cock is slightly opened, the permanent gases which have collected in the top of the condenser will blow out, followed after a time by the ammonia. The permanent gases pass through the small pipe, and escape through the water in the shape of bubbles. When these cease and a sharp crackling is heard, ammonia gas will be escaping and the cock should be shut. Some authorities prefer to extract the air from the top of the liquid-receiver, but, though to a slight extent they are supported by theory, experience points to the condenser as being the proper place. A similar operation is gone through in extracting air from a plant using submerged condensers.

**Cleaning Condenser Coils.**—In order to get the best heat exchange, condenser coils should be painted black with some anti-fouling paint, and should be kept thoroughly clean and free from rust. The distributing pipe should be cleaned out at very frequent intervals, and an even flow of water over the coils maintained.[1] If the condensers are foul, the heat interchange is retarded and the condensing

[1] See also Liquor-cooling, page 374.

pressure rises. Atmospheric condensers are comparatively easy to keep clean. Submerged condensers are more difficult, as scale and mud are apt to collect on the coils. A method, said to have been used in the United States with some success, consists of putting a solution, consisting of one part of muriatic or hydrochloric acid to two parts of water, into the condenser casing and leaving it until effervescence ceases. After treatment the scale could be washed away as mud. During treatment a considerable quantity of gas is evolved. The same method is commonly employed for cleaning scale out of water-pipes, etc.

**Size of Condensers.**—The actual amount of surface necessary in any individual case depends entirely on the amount of cooling water available. In England the maximum temperature of water varies from 70 to 75°, corresponding, on a change of 10, to a pressure in the condensers of 154 to 168 lbs. per square inch. In other countries the temperature of the water is often higher. The condenser must, therefore, be designed to suit the conditions existing. As a general rule, from 40 to 50 square feet of surface per ton ice-making is correct for atmospheric condensers, and 50 to 60 square feet for submerged condensers; the larger the machine the less is the surface per ton necessary. In any case the condenser should have sufficient surface to suit the maximum capacity to be expected of the refrigerating machinery.

**Water Required.**—The amount of water required likewise varies with the temperature. There is no advantage in passing an unlimited amount of water over the condenser, as under these conditions a considerable portion of it fails to come into contact with the pipes and is useless. The best conditions exist when about 5 lbs. of water per foot-run per minute pass over the condenser and run down over the coils in a thin film. At this rate each pipe should heat the water some $2\frac{1}{2}°$. If feathers are omitted, it will be found that the water will fall in streams, or, if there is a breeze, will leave the condenser altogether. In this case the top pipe would heat the water from 2 or 3° and the succeeding pipes hardly more than 1°. With well-water at a temperature of 50 to 55° about $2\frac{1}{2}$ gallons per ton ice-making per minute are required.

**Supplementary Condenser.**—In some cases where water is scarce, a supplementary or fore-condenser is employed. This consists of a coil of pipes through which the gas passes on its way to the condensers proper. In this it is cooled by the waste water from the condensers, which passes over the coils after it has left the condensers.

The relation the subsidiary condenser bears to the main condensers is very much the same as that which a fore-cooler bears to an ice-tank.

**Cross Connections.**—Where two machines are employed, they should be cross-connected on the delivery side in such a manner that, if necessary, either machine can be worked with either condenser alone, or both machines can be worked with both condensers together. This is exceedingly useful in cases where one condenser is shut off for repair, or in case of accident.

## The Expansion Side.

The expansion side of a refrigerating machine is quite the most simple portion of the apparatus. The arrangement of the expansion pipes depends very much on the objects to be attained and the refrigerating agent in use. With ammonia and direct expansion these consist of coils of pipes—usually 2-inch—placed in the rooms to be cooled, and, when the temperature is below freezing, generally in

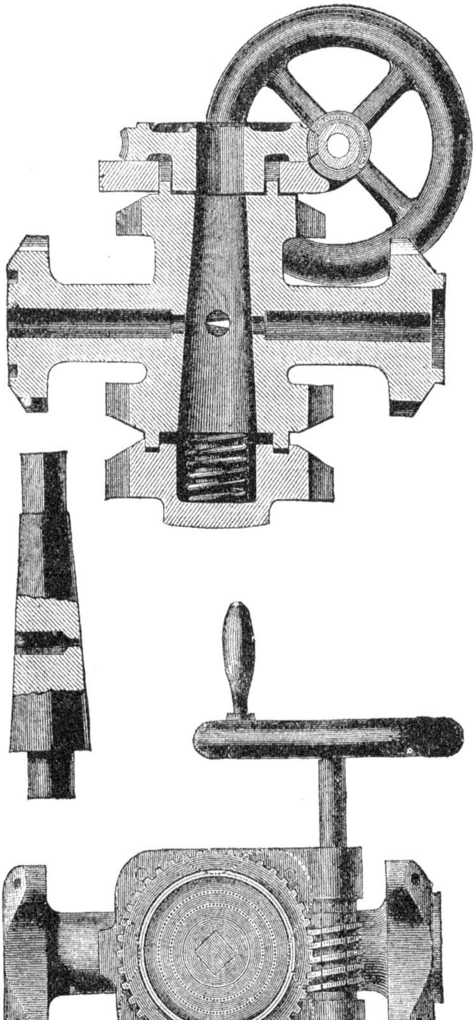

Fig. 40.

a ratio of 1-foot run of 2-inch piping to 12 cubic feet of space. When cooling brine, either for brine circulation or for ice-making, the expansion coils are usually placed in a tank and surrounded by the brine. The arrangement, however, can be either after the pattern of the submerged condenser, the Pontifex cooler, fig. 101, or of the Baudelot cooler shown in fig. 110 (Brewery Refrigeration), and can and does vary largely in practice. With $CO_2$, brine-coolers of the submerged type are most largely used. There are two methods of feeding the refrigerant into the coils, known as top and bottom feed. These are referred to at greater length in the chapter on ice-making, page 279.

**The Expansion Valve.**—The refrigerant is introduced into the expansion coils through a valve which is capable of very fine adjustment, such as that shown in fig. 40.[1] It will be noticed from the section of the plug that the circular opening does not pass right through the valve, but leaves a thin metal bridge in which an opening in the shape of a sharp-pointed V is left—this is shown in the top section of the valve—the plug is turned by the worm and worm-wheel, and the narrow end of the V is the first to open. The worm-wheel is graduated, and shows by its position the extent to which the valve is open, and very fine adjustments can accordingly be made. A valve of somewhat different design, but in which the first discharge is also made through a V, is shown in fig. 41[2]; the drawing is largely self-explanatory, and the same remark applies to fig. 42,[3] which is a combined stop-cock and expansion valve, the idea being that if at any time

[1] Messrs L. Sterne & Company, Ltd., Glasgow.
[2] Messrs the Frick Company, Waynesboro, Pa., U.S.A.
[3] Messrs the Triumph Ice Machine Company, Cincinnati, O., U.S.A.

the expansion valve requires to be repaired, the plug containing it can be shut off and the expansion-valve

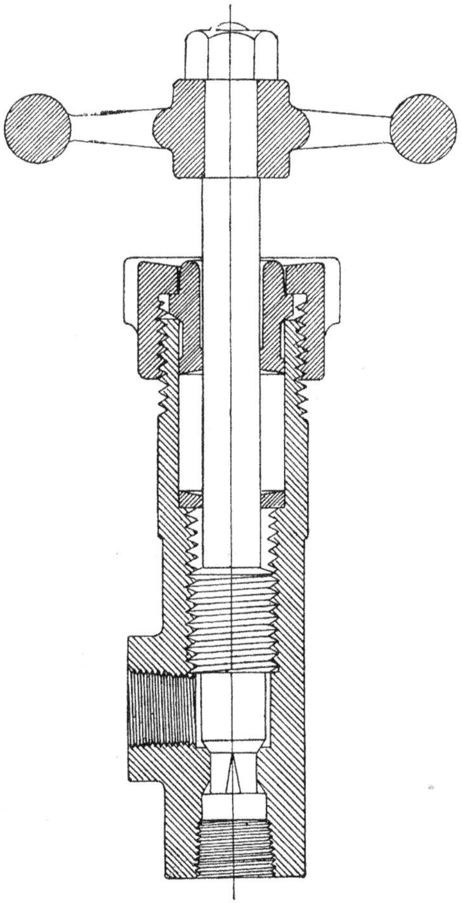

Fig. 41.

spindle removed, its place being taken by a plug which is issued with each valve.

**Stop Valves.**—There is no part of a refrigerating plant which gives more trouble than the valves. This is

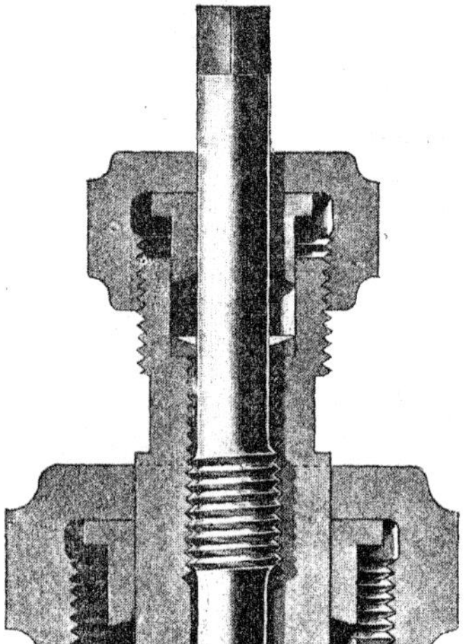

Fig. 42.

particularly the case with ammonia, which is peculiarly searching. No copper or brass can, of course, be used, and steel is the material usually employed. In some cases the valves are made with nickle or with soft metal seats, but steel to steel seems to give the best results. As a matter of convenience, valves are better than cocks, as they can be more readily opened and closed in cases of emergency. It is, however, of great importance that the valves should be tight, especially when any portion of the system is being pumped out and opened up for repairs, as a leaky valve means a loss of ammonia.

Fig. 43[1] illustrates two De la Vergne stop-cocks. The top figure is a 2½-inch cock, and the lower one a 1-inch cock. The peculiar construction will be noticed; both the top and bottom of the plug is covered with a cast-iron cap. This makes a gas-tight joint with the valve body, which is seated on a lead gasket, filling the annular recess in the cap. These caps are provided to prevent any leakage of gas; that covering the plug is removed when the valve is operated, and is replaced subsequently. A spring is inserted between the lower cap and the valve plug, with the object of keeping the plug up to its seat and preventing grit getting in between the metal and scoring the surfaces. Fig. 44[2] illustrates an ammonia valve with double seats. The valve spindle has a right- and left-handed thread, and the valve seat is made top and bottom on white metal. Each maker has his own design of valves and fittings, and these designs naturally vary very largely; the accompanying illustrations are merely given as examples of the principles aimed at. This remark also applies to pipe unions and joints, a great variety of which are employed. The ideal joint is a welded one, or no joint at all, and in these days of the almost universal use of electricity it is

[1] Messrs L. Sterne & Company, Ltd., Glasgow.

[2] Messrs the Triumph Ice Machine Company, Cincinnati, O., U.S.A.

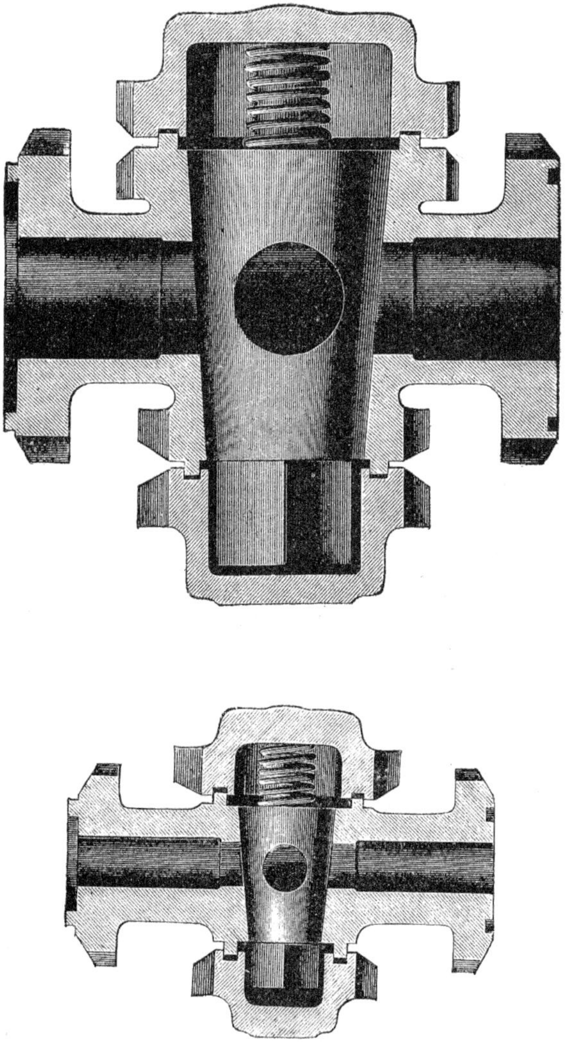

Fig. 43.

to be hoped that this joint will come into more general use. Every joint in a plant is a weak place and a possible source of trouble, and anything that can minimize their number is an advantage. Where joints must occur, the usual practice in some form or other is to provide a screwed

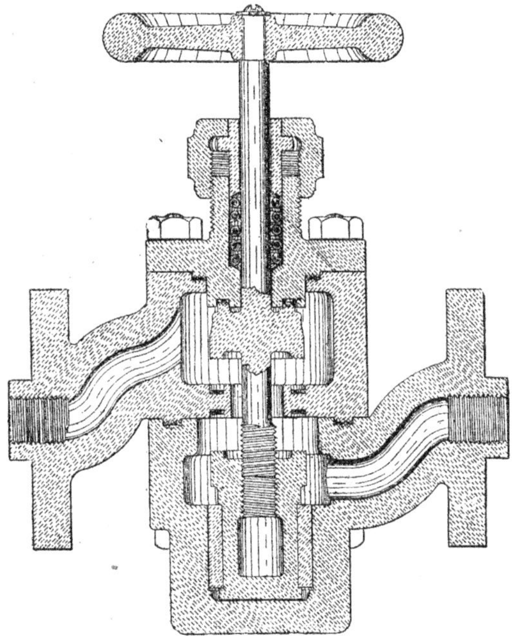

FIG. 44.

and soldered joint; this is obtained by screwing the pipe into a flange having an annular recess on its outer face. After the pipe is screwed home, this recess is carefully filled with solder. The parts of the pipe and flange in contact with the solder must, of course, be first tinned. The joint between the flanges is made with a lead washer.

**The Ammonia Dryer.**—The dryer is a very useful fitting, which is frequently omitted from plants. Ammonia has a great affinity for water, and will readily absorb it; at the same time, ammonia, to give the best results as a refrigerant, must be absolutely anhydrous. In its preparation ammonia is passed over quicklime or caustic soda in order to absorb any moisture it may contain, and as there is always a chance of its picking up water after it has been put into a plant, the dryer—a vessel filled with similar material—is provided and placed as a byepass on the suction side. This vessel is fitted with trays containing, as a rule, quicklime, and if moisture is suspected in the plant—its presence can easily be ascertained—the suction gas is passed through the dryer, and passes away in an anhydrous condition. When the lime is exhausted by the moisture absorbed, the dryer is pumped out, the cover is removed, and a fresh charge of lime is inserted. The apparatus is cheap, occupies little space, and is an exceedingly valuable adjunct to any plant.

# CHAPTER VII.

## AUXILIARY PLANT.

**Buying Machinery.**—It is presumed that the capitalist or business man who desires to buy machinery for producing a certain effect or carrying out a definite purpose, be it the production of power, refrigeration, or electric light, will confine himself and his knowledge to the results he wishes that machinery to give him ; that is to say, he will regard the machinery as only an indispensable accessory to his object, not the object itself. He will thus be in the position of wanting something done in order that he may do other things in which he is more particularly interested, and from which he hopes to attain monetary gain. Suppose, for instance, he is anxious to erect a cold store or an ice factory. His object in doing this will be the derivation of a profit from the storage of perishable articles or from the sale of ice. He will know that for both purposes refrigerating and other machinery is as indispensable as the buildings in which the plant is situated, but he will mainly interest himself in selling his ice and securing goods with which to fill his cold stores. These will be his chief care, and on them he will lavish his skill and business acumen. As it is not his business, he will probably know little or nothing about the details of the machinery he wishes to buy, but, in order that his profits may be large, he will desire that it shall be capable of cooling his stores, or making his ice, at as small a cost as possible. He will, however, recognize that the cheapness of an article depends on more things than price, and will clearly discriminate between expenditure on capital and revenue account, and, while endeavouring to keep the former within reasonable limits compatible with the provision of first-class work, he will be anxious that the latter should be kept at a minimum. Having clearly made up his mind what he wants to do, he will turn his attention to the best means of doing it, and will adopt one of two courses. Either he will go to some manufacturer and himself obtain a quotation for the whole plant, give the orders, and attend to the details, or he will consult some independent expert in whom he has confidence, who must essentially be a good business man with a knowledge of the markets, and who will advise him and tell him how, having regard to all the circumstances, he can best and most cheaply attain the objects he has in view. In either case he will be guided by the advice he receives. The secret of buying machinery cheaply is to buy it individually, each machine from its own maker. The expert, if he is wise, will follow this course, and, after designing his arrangements and finding out definitely and exactly what results he wishes the machinery to give, will not attempt to design that machinery itself, but will draw his specifications, and call for competitive tenders, from the various manufacturers for machinery of standard design in their respective works. In this specification, he might mention details which he would prefer to have if they could be obtained at a reasonable cost, but, at the same time, it should be clear that machinery of standard size and make is wanted, and that the non-provision of these details should not disqualify any contractor from tendering.

The manufacturer has devoted his time, experience, and money to finding out what design is, from every point of view, the best and most economical. He has made this his standard, and he has all the patterns and templates necessary for quickly making the various parts. He knows exactly what a machine of a certain standard size will cost him to make, and what results it will give under certain specified conditions. He also knows that his success depends on his making these standard machines and these results as good as, or better than, those of his competitors, and he is prepared to guarantee that they will do what he says. Until recently the curse of the English engineering trade has been the meddlesome way in which engineers who have been buying machinery have interfered with the makers' designs, and have either insisted on having the machines built from their own design, or at least in specifying the leading dimensions. The result of this has been that makers have had to make, and the client to pay for, new designs and patterns for every machine ordered. The contractors have never had a chance of standardizing their work, and consequently have had to quote high prices and excessive time for delivery. It is a matter of comparative indifference to the business man in what way the details of his machinery are arranged, provided that it is fully capable of performing the work he requires of it, at least as economically in all respects as that made by any other maker. The expert should recognize this, and, while satisfying himself that he is getting what he wants and maintaining a general, and sometimes very necessary, supervision over the quality of the materials used and the workmanship employed in the manufacture of the machinery, he should devote himself and his experience to the general design of the plant and buildings, of which the machinery forms only a part, and with which the various manufacturers of the machinery have no concern. The object of an expert is to save his client money, not to exercise his idiosyncrasies; and he cannot hope to do this unless he fits his designs to the use of machinery, the standard design of which is the very essence of its cheapness and efficiency.

## ENGINES.

The steam-engine invariably forms part of the refrigerating machine, and is supplied with it, each maker having his own standard. There is not much room, therefore, for choice apart from that of the refrigerating machine, for when the one is decided upon, so also to a great extent is the other. Since the days of the low-pressure single-cylinder type of engine, however, great improvements have been made, and the steam-engine is now, both in construction and efficiency, vastly superior to what it used to be. These improvements can be traced largely to the use of high-pressure steam, superheating, multiple expansion, and the use of condensers, as well as to improvements in valves and valve gear, the introduction of piston valves and Corliss, and other quick-release trap valves. Constant and steady running has also been rendered possible by improvements in lubricants and methods of lubricating, as well as in adjustable bearings and anti-friction metals.

The principal points to be considered about the engine, as indeed about the whole machine, are:—

**Efficiency.**—By this is meant not merely efficiency under test, but under ordinary working conditions. The engine should not require more than 20 lbs. of steam per I.H.P. hour when working condensing, or 25 lbs. when working non-condensing. These may be taken as outside limits for good work.[1] Owing to the slow speed at which refrigerating plants usually run, the cylinder condensation is rather serious. This may be guarded against by jacketing the cylinders with live steam or superheating, and the majority of the best makers now follow this practice.

**Simplicity.**—The more simple the design of the engine the better, as an excessive number of working parts increases the attention and repairs required. The working parts should be easily accessible for repairs and for examination when running.[2] Many a hot bearing would have been avoided had it been more easy to get at the working part. It is an advantage, too, for the main bearings to be so constructed that they can be got at, and the brasses taken out, without dismantling the machine.

**Durability and Solidity.**—As refrigerating machines may require to run day and night for long periods, solidity in construction is absolutely necessary. The bedplate and frames should be of massive design and well braced. The machine should be firmly bolted down, so that there need be no vibration when it is at work. It is also desirable that the steam-engine should form part of the complete machine. The eyes and pins of all joints that are not fitted with brasses should be case-hardened. Changes of section should be gradual, and corners should be rounded. The shafts and motion bars should be of high quality mild steel. If the machines are of standard

[1] As a matter of fact makers can guarantee a steam consumption of 15 lbs. per I.H.P. hour for compound condensing engines running at 60 revolutions per minute when using steam at 180 lbs. per square inch. If the steam is superheated 100°, 14 lbs. per I.H.P. hour can be obtained.

[2] This, of course, does not apply to high speed enclosed electric light engines; they are not being considered, and in any case work under different conditions.

design, all the working parts should be made strictly to gauge and template, and should be numbered, so that in case of a breakdown, it is only necessary, in order to obtain a duplicate part, to telegraph the number of the part broken to the makers. Similar working parts should be interchangeable from machine to machine, and spare brasses, etc. should be kept for emergencies. When it is remembered that thousands of pounds' worth of goods are liable to be ruined if the engine breaks down and the temperatures rise, it will be seen that a few pounds spent in what are sometimes called refinements, but which enable repairs to be quickly and cheaply executed, are well invested. For the same reason it is always advisable, if possible, to have duplicate machines.

## Valves and Valve Gear.

In the early days when steam pressure did not exceed 50 lbs. in the square inch, the ordinary D slide valve could be used. When steam pressure, however, began to mount the gauge, it was found that the pressure of the steam acting on the whole area of the valve caused an immense amount of friction, and frequently resulted in the breaking of eccentric rods, valve spindles, and similar parts.

There was also another trouble with the old D valves. In engines of large power the ports were of considerable size, and as the travel of the valve was slow, it would take some time to close the port, and would, as a consequence, wiredraw the steam. In order to avoid these disadvantages, attempts were made in the former case to reduce the friction by balancing the valve, and in the latter to prevent wiredrawing by employing some gear that would shut the valve as instantaneously as possible, and so give a quick cut-off.

**Piston Valves** have been largely employed for both these purposes, but more particularly in vertical engines. They have come commonly into use in connection with high-speed vertical engines for electric lighting, but are used by some refrigerating machinery makers, notably Haslam, for the smaller vertical machines. There is nothing particular to describe about a piston valve. It consists of a piston containing ports, through which steam is admitted to the cylinder at the proper time. Its cylindrical shape prevents there being any unbalanced pressure, and also ensures good lubrication.

**The Corliss Valve and Gear.**—The Corliss valve, which is now so much used for the high-pressure cylinders of refrigerating plants, was invented by G. H. Corliss in America in 1848. The object is, in common with all trip gears and slip motion, to prevent the wiredrawing of the steam as well as to give a balanced valve.

The valve gear has undergone extensive modifications since it was first introduced, but the principle remains unchanged. A steam cylinder fitted with Corliss valves is shown in figs. 45 to 47.[1] Four quite independent ports, two at each end of the cylinder, are used for admitting and exhausting the steam. The admission valves are opened against the pressure of a spring, and when the desired point of cut-off has been reached, it is released by trip gear and closed very suddenly by the spring. The springs are connected to and brought to rest by a dash pot, which prevents any shock or jar. The engine is governed by varying the point of cut-off instead of by varying the pressure, as in a throttle governor. The

[1] Messrs Yates & Thom, Blackburn. (Dobson's patent.)

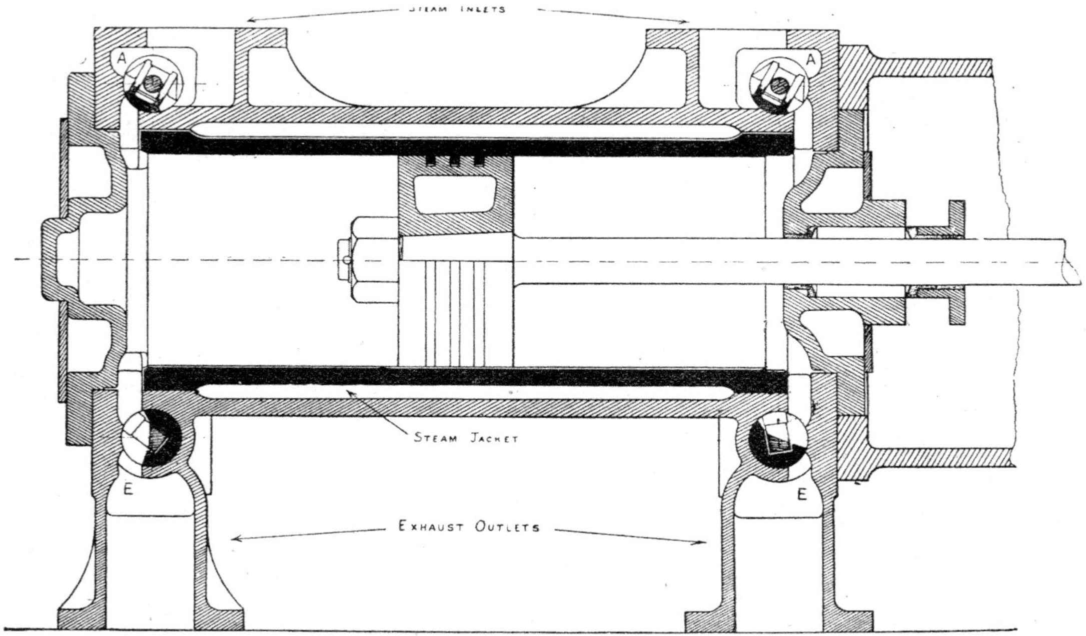


Fig. 45.

valves are balanced, except when the port is closed, so that there is little friction. In the figures the exhaust valves *EE* are connected, as shown, to the exhaust eccentric, and the steam valves *AA* are connected to the trip gear and operated as follows. The primary motion is received

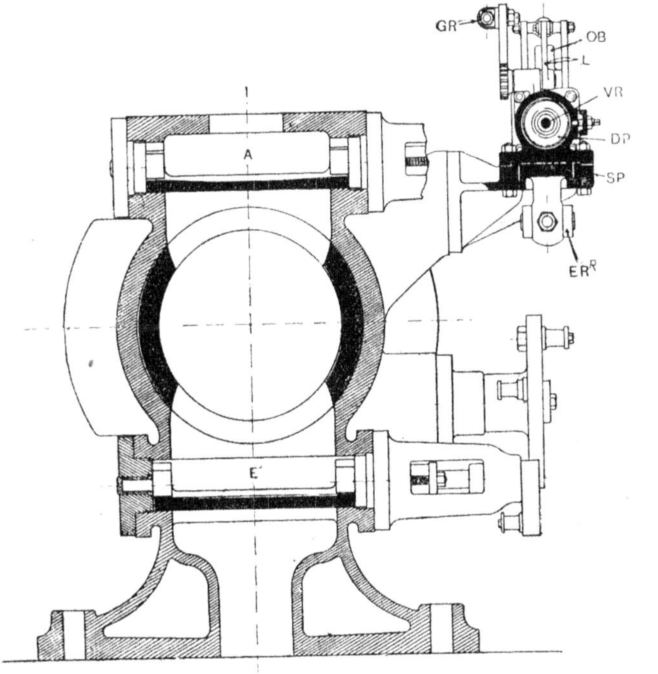


FIG. 46.

from the steam-engine through the eccentric rod *ER*. Attached to this is a slide plate *SP*. This engages with the valve rod *VR* through the catch box *CB*, the drop block *DB*, and the spring *S*. The dash pot, which contains a strong spiral spring, is shown at *DP*. When the engine is at work, and as the slide plate and catch boxes reach the end of the stroke, the drop block *DB* engages

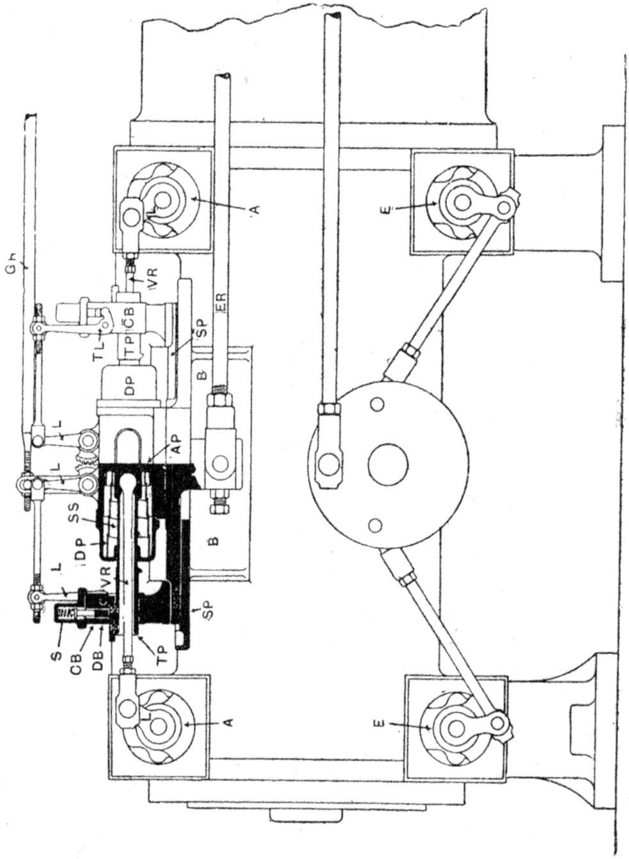


FIG. 47.

with the truck piece *TP*, which is connected to the piston of the dash pot and the valve rod *VR*. As the slide continues to move forward, the truck piece is drawn out against the pressure of the spiral spring, and the valve is opened. When the proper and predetermined point of the cut-off has been reached, the toe lever *TL* raises the drop block *DB*, disengages the truck piece, and releases the valve rod. The spiral spring then acts—the valve rod flies back and the valve is quickly closed. The air compressed at the end of the dash-pot cylinder prevents any jar. The governor acts on the toe lever *TL*, and so determines the point of cut-off.

Figs. 45 and 46 show respectively a transverse section and a longitudinal section through the cylinder and valve gear.

**The Meyer Valve and Gear.** — The Meyer variable expansion gear is probably the best known and most

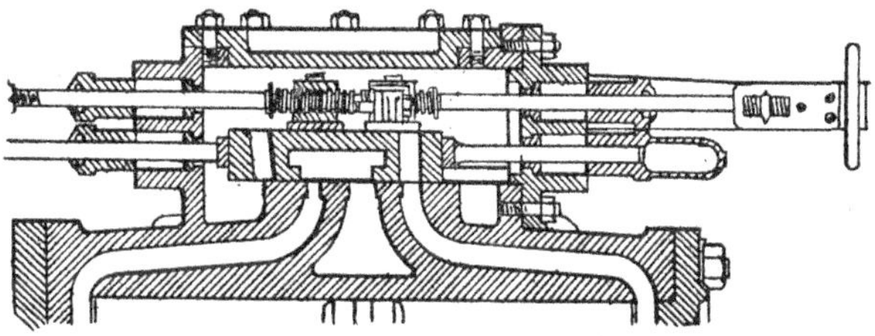

FIG. 48.

largely used valve gear there is. The general design is shown in fig. 48. It is worked by two eccentrics, and consists of two separate valves working independently of each other. The main valve is double-faced, and the steam is admitted to the cylinder, not at the extreme edges of the valve as is usual, but through two ports cut in it and corresponding in width to the ports in the cylinder face. This main valve is driven by one of the eccentrics in the ordinary way. The Meyer slides work on the back of this valve as shown, and they cut off the steam at any desired point by closing the ports of the main valve beneath them. These valves are worked by their own eccentric, and the variation is obtained by means of right- and left-handed screws on the valve spindle. The end of this spindle passes through a gland in the valve chest and is formed square. By this means

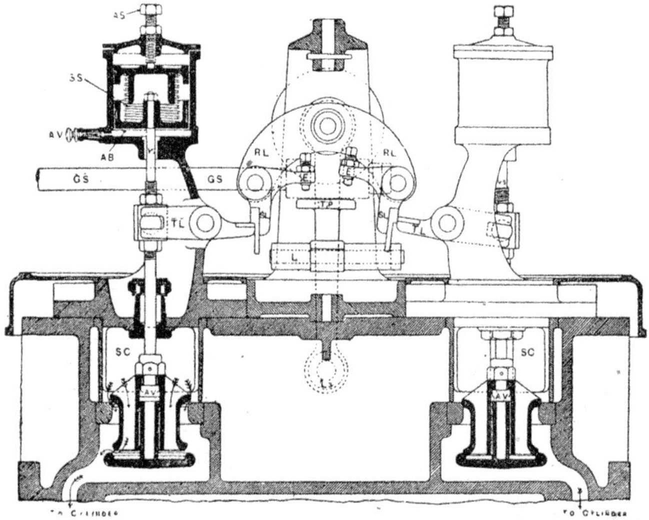

FIG. 49.

it can be engaged by the square on the hand-wheel and rotated while the engine is working. By turning the valve spindle, the distance between the Meyer valves can be altered, and the cut-off varied without any variation in the travel.

In addition to the convenience of being able to alter the cut-off in this way, the Meyer gear, owing to the fact that the two valves have combined movements in opposite directions, gives a much quicker cut-off than the ordinary single eccentric motion.

**The Proell Valve and Gear.**—Figs. 49 and 50 illustrate the Proell valves and gear. It consists of two double beat admission valves *AV* placed one at each end of the cylinder. The variable cut-off is obtained by the action of the governor, which communicates its rotary action through levers to the vertical spindle of the two double beat valves. The valve spindles *US* are alternately

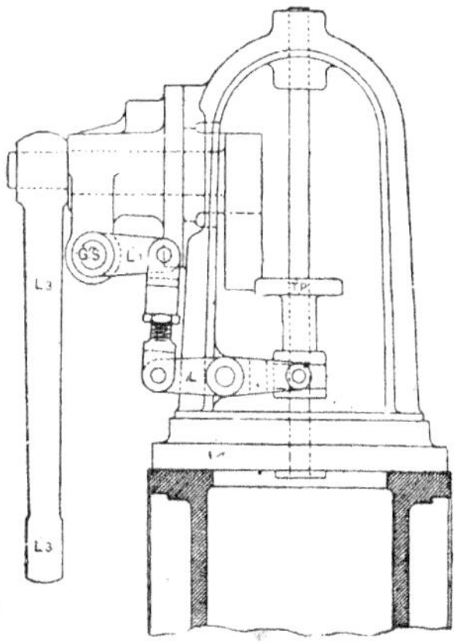

FIG. 50.

lifted by the trip levers *TL*. These levers are in contact with the slip gear at the end of the slip lever *SL*.

These slip levers are actuated by the rocking lever *RL*, which gets its motion from the eccentric rod attached to the lever $L_3$. The inner arms of the slip levers are fitted with special steel set screws, which engage at each stroke with the trip pad *TP*. The position of this trip pad on the spindle, and therefore the movement of cut-off, is controlled by the governor, which acts through the governor shaft *GS*, as shown in fig. 50. As soon as the inner arm of the lever *SL* comes in contact with the trip pad, any further movement of the rocking lever necessarily disengages the trip gear and allows the valve spindle to fall. The fall is accelerated by the action of a spiral spring *SS* on the top of each valve spindle, and jar is prevented by the dash box *AB*. The air pressure in this dash pot may be so adjusted by the valve *AV*, that the valve falls on its seat without shock.

Other valve gears, such as Rider's cylindrical trapezium, variable expansion valves, and Weelock valves, have at times been used for refrigerating machinery, but they are not so common as the types just described, and need not be illustrated or described.

## Boilers and Steam-Pipes.

The selection of suitable boilers for cold storage and ice-making plants is, like that for other factories, largely a matter of convenience. Each particular form has its staunch advocates, and its claims are urged often quite irrespectively of the conditions under which it has to work. The types may be conveniently divided into three classes:—

1. The Cornish and Lancashire boiler.
2. The marine dryback and locomotive boiler.
3. The water-tube boiler.

**The Cornish and Lancashire Boiler.**—It may be accepted as a safe rule that for all-round work, and where the necessary space is available, the Lancashire boiler is the most suitable and economical for the requirements of an ice factory or cold store, where steam has to be kept up continuously day and night. These boilers are simple in construction and have ample internal space. They are easy to clean and examine, and, owing to the amount of water they contain, they possess great steaming capacity. For the same reason, and in order to avoid stressing the boiler too heavily by unequal expansion, steam should be raised very gradually. This, though a drawback in cases where the boilers are liable to meet heavy and unexpected demands on their steam, is not a matter of importance where the load remains as constant as it usually does in refrigerating work. The Lancashire boiler is merely a development of the Cornish type. The general principles are the same, but the former has two fire flues, while the latter has only one. In the old days the flat end plates of these boilers were stayed by longitudinal stays running from end to end of the boiler shell. Owing, however, to unequal expansion, these invariably give trouble, and though some engineers still insist on their use, the better practice is to use gusset stays, which are not open to the same objection. These stays are formed of flat boiler plate, and are riveted to the shell of the boiler and to the end plates. There has also been for some years a difference of opinion as to the flanging of the end plates. Some engineers insist on both the front and back end plates being flanged inwards and riveted to the shell, but the more usual practice is to adopt this course for the back end plate only, the front end plate being attached to the shell by an angle iron.[1] This method is equally safe, less expensive, gives more room for the fittings, and has a much better appearance than the flanged front plate.

[1] Recently Messrs Danks of Oldbury have built some boilers, the front end plates of which are flanged outwards and attached to the shell and the fire flues by a double row of rivets, all of which are exposed to view. This construction makes it particularly easy to lagg the fronts of the boilers with non-conducting composition.

**The Marine, Dryback, and Locomotive Boiler.**—Though these boilers are sometimes put in where it is necessary to economize floor space, or where the water-tube boiler is for some reason or other objected to, they are not recommended except for small powers, as they are not economical and, owing to the large number of tubes, they are very difficult to clean and examine. This objection particularly applies to the locomotive type. With the dryback or marine type the saving in space is the great consideration.

**Water-tube Boiler.**—The same consideration as to space applies to the water-tube boiler, one of which is illustrated in fig. 51.[1] The boiler is hung from girders, and the tubes are free to expand and contract with the varying temperatures. There is no likelihood, therefore, of leakage or fracture occurring. As the water is inside the tubes, the scale can be got at by means of a scraper; nevertheless, cleaning is apt to be a tedious and expensive process. Water-tube boilers are frequently fitted with superheaters capable of giving from 100 to 150° of superheat to the steam, and as these boilers, if forced, are liable to produce wet steam, this is an adjunct which should not be omitted.

**Boiler Seatings and Floor Space.**—The expense of the necessary seatings is an item which is sometimes lost sight of when the type of boiler to be installed is under consideration. The Lancashire boiler requires expensive seatings, the locomotive and marine type hardly require any at all, while those for the water-tube type are comparatively inexpensive. As illustrating the difference in floor space occupied by a Lancashire boiler and a water-tube boiler respectively, a 200-I.H.P. Lancashire boiler capable of evaporating 4200 lbs. of water per hour would measure

[1] Messrs Babcock & Wilcox, Ltd., London.

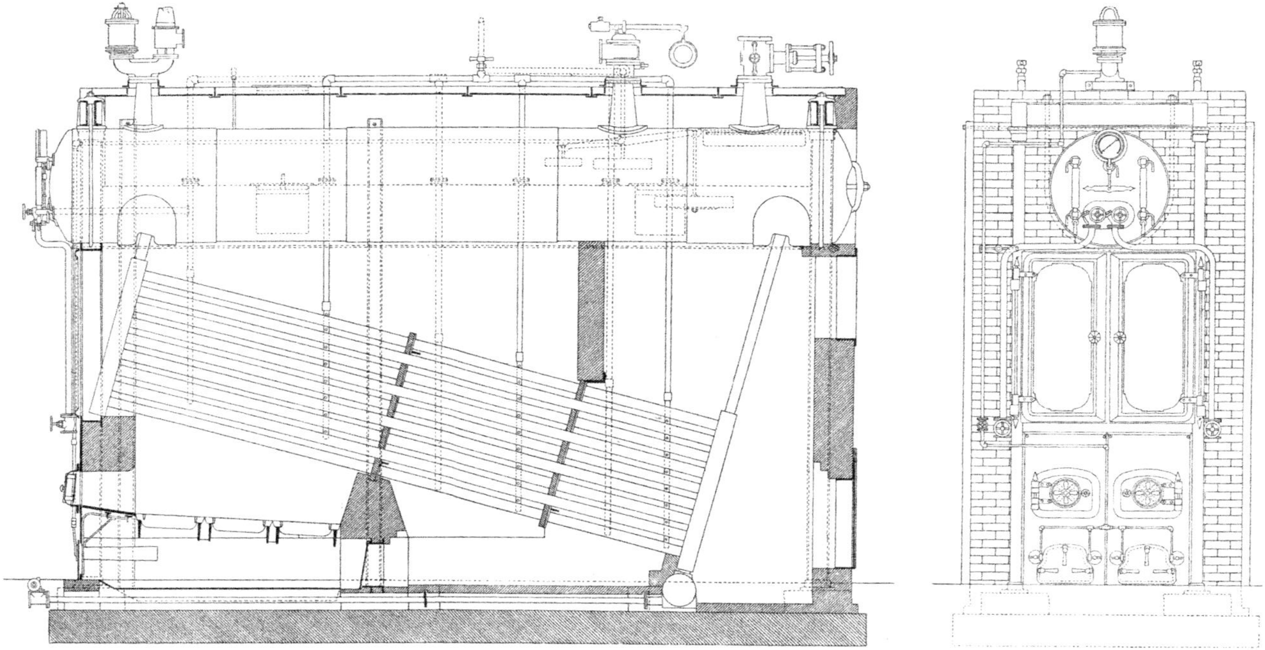

FIG. 51.

To face page 12.

26 ft. long by 6 ft. 6 in. diameter, and with its seating would require a floor space 10 ft. 3 in. wide and 28 ft. 9 in. long; while a 200-I.H.P. water-tube boiler would only require, with its seating, a space 22 ft. long by 6 ft. 6 in. wide, the flue to the chimney being omitted in both cases.

**Boiler Fittings.**—Opinion varies as to what are or are not indispensable fittings on a boiler, but two safety valves—one a dead weight, and the other a lever or spring valve—should be provided. Sometimes a high- and low-water alarm valve is fitted. The object of this valve is, by the blowing of a whistle, to attract attention should the water-level in the boiler, through carelessness or accident rise too high or fall too low. Where a number of boilers are fitted in a battery, an isolating, or non-return, valve should be fitted between each boiler and the common steam main; this not only prevents the possibility of men being scalded to death by the accidental opening of the steam valve should one of the boilers be under repair, but ensures that each boiler does its fair share of work.

**The Prevention of Explosions and Inspection of Boilers.**—Though undoubtedly many explosions which have occurred in the past can be traced to bad workmanship, still, should a boiler, however well made, be badly handled and not properly cleaned, a risk of accident is sure to be incurred. One of the most fruitful causes of damage to boilers is the severe racking and straining they receive when the shell is too suddenly heated or cooled. Fusible plugs should always be provided on the top of the boiler flue. These prevent overheating, should the water get too low, or the flue become covered with an oily deposit from the cylinder of the engine when condensed steam is used,[1] as well as from an excessive accumulation of scale. In order to prevent explosions, boilers should be regularly inspected and cleaned. Too much care cannot be given to this matter, and boilers should be laid off at regular intervals of time to enable cleaning to be done. It is also advisable to insure boilers in one of the Boiler Insurance Companies who have a staff of well-trained inspectors.

**Fuel Economizers.**—In order to reduce the coal bill it is of importance to make use of as much of the heat of combustion as possible. This is the object of the economizer, which consists of a nest of tubes built into a bye-pass flue, so that the hot gases on their way from the boiler to the shaft can pass around them. The feed-water is pumped through these tubes on its way to the boiler, and its temperature is raised, so that it enters the boiler at or near boiling-point. The tubes are kept free from soot by scrapers actuated by a small steam-engine or electric motor. Consideration of space, however, sometimes necessitates this useful apparatus being omitted. Another type of feed-water heater makes use of the heat contained in the exhaust steam; these are well known and need not be described.

Besides utilizing heat which would otherwise go to waste, a feed-water heater is useful inasmuch as, by heating the water before it enters the boiler, lime and other substances held in solution are deposited in the heater instead of in the boiler, where they would form scale. This is dealt with more fully under the head of water-softening.

[1] There are numerous ways of getting rid of this oil, as, for instance, by the use of oil-filters. A little sapon added to the boiler feed-water has a good effect.

**Automatic Stokers.**—In order to get the best results out of the coal, it should be continuously and gradually fed on to the furnace, not thrown on in large quantities at a time, as is sometimes done with bad hand-stoking. There are three kinds of automatic stokers—those which coke the coal before consuming it, those which sprinkle the coal over the whole area of the furnace, and the underfed stoker. In the two first kinds the coal is fed into hoppers either automatically or by hand, and is then fed into the furnace. The fire-bars are kept moving by gearing driven by a small steam-engine, and they feed the coal gradually on towards the back of the furnace. Automatic stokers are of importance in cities, as they are, or can be made, smokeless.

**Steam-Pipes.**—As steam pressures tend to creep up, more and more attention has to be given to steam-pipes and fittings. In the old days cast-iron was considered good enough, and for low pressure it undoubtedly serves the purpose. For high pressures, however, it is not safe, and some nasty accidents have occurred owing to its use. Steel pipes with malleable iron or cast-steel flanges welded on, or screwed and brazed, are perfectly safe, and should be used.

In order to avoid condensation in the pipes, the engine-room and boiler-house should be so placed as to render the run of steam-pipes as short as possible. It is, perhaps, hardly realized by power-users how much heat is wasted by radiation from steam-pipes and boiler tops, and these and all the flanges should be covered with non-conducting composition. As lagging can be made removable, there is no reason at all why flanges should not be covered. Left uncovered, they are responsible for a great deal of waste.

Figs. 52[1] and 53 show a sectional steam-pipe covering which is moulded to fit the flanges and pipes, and is remov-

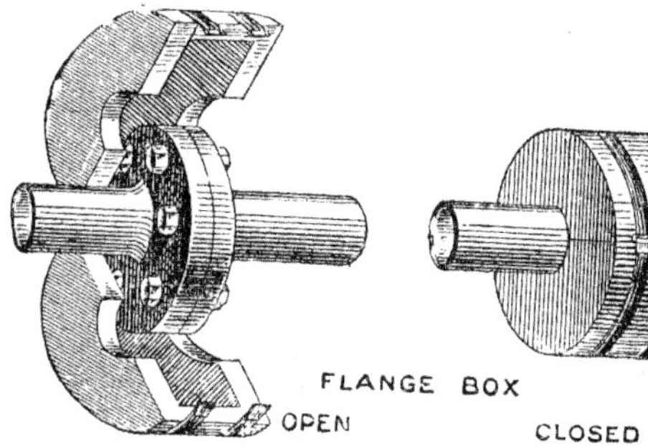

FIG. 52.

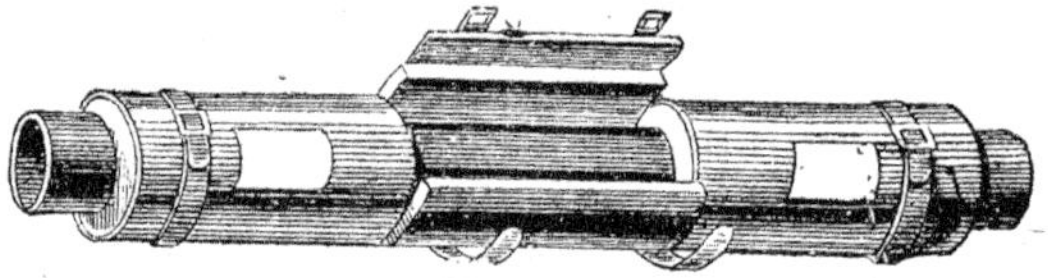
FIG. 53.

able at will by slackening the brass bands. The same firm also makes a flange joint, fig. 54, which consists of a

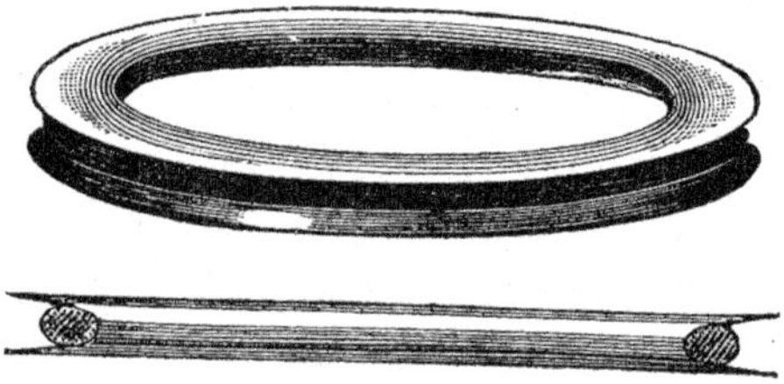
FIG. 54.

double ring of soft metal grooved outside, the inside being filled with asbestos or hemp. The rings are made

[1] Messrs A. Haacke & Co., Homerton, London.

of copper, lead, zinc, white metal, or steel, and readily adjust themselves to any unevenness of surface. They have the advantage that they can be used again if the joint is broken.

**Water-softening.**—The bulk of the water used for boiler-feed and ice-making purposes contains certain proportions of lime and magnesia. These are principally carbonates and sulphates, less frequently chlorides and nitrates. Carbonate of lime, the most common, owes its presence in water to the action of dissolved carbonic acid. Distilled water, in which there is no carbonic acid, can only hold 1·3 grains of carbonate of lime in solution per gallon. If, however, water contains much carbonic acid, it can hold as much as 60 to 70 grains in solution.

When water heavily charged with this carbonate is boiled, the carbonic acid is driven off and the whole of the carbonate, except the 1·3 grains above mentioned, is precipitated in the boiler or vessel and forms a more or less soft incrustation or scale. The removal of this carbonate without boiling has been successfully attained by the Clark process, which consists in adding lime to the water. This lime absorbs the dissolved carbonic acid and forms carbonate of lime, which precipitates. The carbonate previously held in solution, being deprived of the carbonic acid which supported it, also precipitates, and, with the exception of the small proportion of carbonate which remains permanently dissolved, the water is left perfectly clear and soft.

When carbonate of magnesia is present, this simple process will not suffice, as the removal of the carbonic acid will not, in itself, cause the precipitation of the magnesia. To do this an excess of lime has to be added to the water. This decomposes the carbonate of magnesia and converts it into hydrate of magnesia, which is insoluble in water. Sulphate of lime is an exceedingly troublesome constituent of boiler feed-water, as it forms a particularly hard crystalline scale, which is very difficult to remove. Its solubility in water does not depend on the presence of carbonic acid, and it is not precipitated when the carbonic acid is removed by the addition of lime. Precipitation, however, can be brought about by the addition of carbonate of soda, which decomposes the sulphate of lime, precipitates the resulting carbonate, and leaves in its place a harmless solution of sulphate of soda.

It will be seen from the above that the process of softening water consists in removing first the carbonate of lime and magnesia which are precipitated almost entirely out of the water, and secondly by the removal of those more insoluble salts and sulphates which are removed from the water by the action of another salt which remains in the water, but which is of a harmless nature unless it is allowed to concentrate to an undue extent. This danger, slight as it is, may be entirely obviated by blowing off the boiler and introducing fresh water from time to time.

In the Archbutt Deeley[1] process of water-softening all the operations required consist in preparing the reagent and mixing it with the water to be softened, the separation of the precipitate, and the carbonating of the softened water. The necessary apparatus is exceedingly simple, and, when from 5000 to 10,000 gallons per hour are being treated, consists of a cast-iron tank divided into two equal portions by a steel or wrought-iron partition. These two portions are fitted up in duplicate, the softened water being drawn off from one while the process of softening is at work in

[1] *Proceedings of the I. Mech. E.*, 1898, No. 3.

the other. For smaller plants up to 3000 gallons per hour, only one softening tank and a storage tank are necessary. Figs. 55–57 illustrate the apparatus.[1] The hard water is admitted to the tank through the pipe (1), and the tank is filled to the level of a guage-mark inside it. While the water is being admitted, quicklime and carbonate of soda, in proportions which depend on the character of the water, are mixed with water and boiled

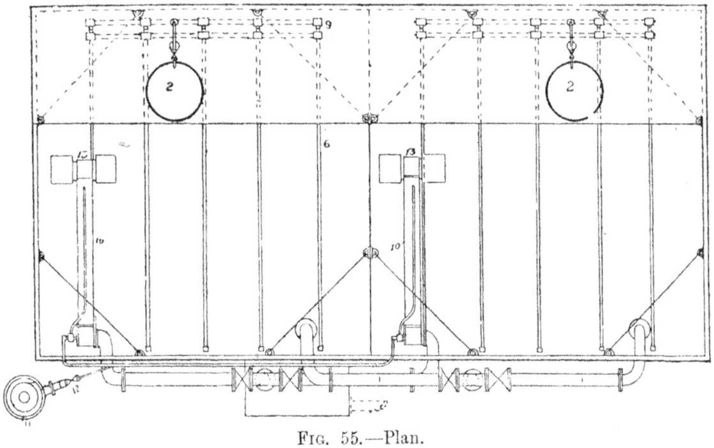

FIG. 55.—Plan.

up by live steam in the small tank (2) placed on top of the main tanks. When the tank is full, the supply is shut off and a circulation of the water is started by the steam-blower (3). By this means water is drawn up through the rose (4), passes through the three-way cock (5), where it comes in contact with a small stream of the prepared chemical solution introduced through the cock (7), and passes back into the tank through perforations in the upper side of

[1] Messrs. Mather & Platt, Ltd., Manchester.

the horizontal pipes (6), and is intimately mixed with the water in the tanks. When the solution has all passed into the tank, the settlement of the new precipitate is accelerated by stirring up the old precipitate left in the tank from previous operations. The particles of the old precipitate attract the particles of the new precipitate and carry them down rapidly to the bottom of the tank. For this purpose the air is admitted to the blower through

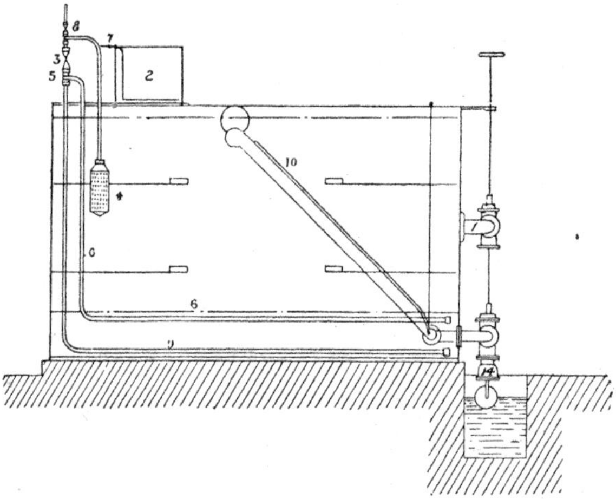

FIG. 56.—Section.

the cock (8), and, by reversing the three-way cock (5), is blown down the vertical pipe and enters the tank by perforations on the underside of the lower horizontal pipes (9). After this agitation has been kept up for some ten minutes the air is turned off. In about an hour's time the precipitate will have settled and the water can be drawn off from the surface through the floating arm.

It has been found that uncarbonated softened water is

very liable to form a troublesome deposit in the feed-pipes, injectors, etc., and an important point in connection

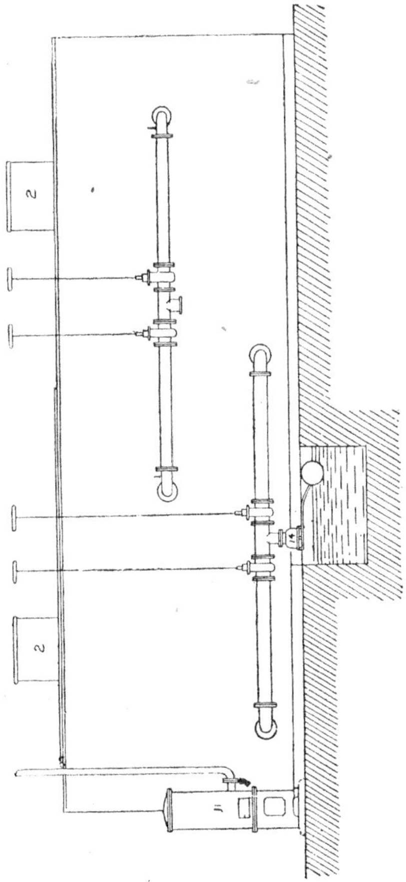


FIG. 57.—Elevation.

with the apparatus is the introduction of carbonic acid into the softened water as it is drawn off. The quantity required is small, and is only just sufficient to convert the free caustic alkali and carbonate of soda into bicarbonates, in which form they are harmless. This carbonic acid is obtained from the coke stove (11), and is, of course, carried over from the boiler with the steam. In distilled water plants it is again driven off in the reboiler.

The costs of softening the water by this apparatus have been given by Mr Archbutt in the paper before referred to as follows:—

| Nature of Hardness. | Cost per 1000 gallons. | | |
|---|---|---|---|
| | Chemicals. | Labour, Interest, etc. | Total. |
| | d. | d. | d. |
| All temporary, . | 0·22 | 0·5 | 0·72 |
| Half temporary, half permanent, . . | 1·06 | 0·5 | 1·56 |
| All permanent, . | 1·90 | 0·5 | 2·40 |

Temporary hardness is the hardness due to carbonate of lime and magnesia, the removal of which is comparatively easy.

Permanent hardness is that due to sulphates and other compounds of lime and magnesia, whose removal presents greater difficulties.

One degree of hardness is caused by one grain of carbonate of lime dissolved in one gallon of water.

## CONDENSERS AND FEED-WATER FILTERS.

The addition of a condenser to a non-condensing steam-engine effects economy in either of two ways:—

1. By the saving of steam due to a shortened cut-off if the power of the engine is required to be kept constant; or
2. A large increase in power if the cut-off is kept constant.

The diagram fig. 58[1] shows the percentage of power that may be gained by attaching a condenser to a non-condensing engine, assuming that the vacuum is 12 lbs. per square inch. The diagram also shows the mean pressure in the cylinder for a given initial pressure and cut-off, clearance and compression not considered. The pressures given in the diagram are absolute.

To find the mean effective pressure produced in an engine cylinder with 90 lbs. gauge (=105 lbs. absolute) pressure, cutting off at ¼ stroke, find 105 in the left-hand vertical column, follow the horizontal line to the right until it intersects the oblique line that corresponds to the ¼ cut-off, and read the mean total pressure from the row of figures directly above the point of intersection, which in this case is 63 lbs. From this subtract the mean absolute back pressure (say 3 lbs. condensing, 15 lbs. non-condensing) to obtain the mean effective pressure, which in this case for a non-condensing engine gives 48 lbs. To find the percentage of power gained by using a condenser with this engine, read on the lower scale the figures corresponding in position to 48 lbs. in the upper row—in this case 25 per cent. It will be seen that, as the diagram

[1] Kent's *Mechanical Engineers' Pocket-book*, third edition.

does not take clearance or compression into consideration, the results are only approximate.

It is easy to understand that this power gained may,

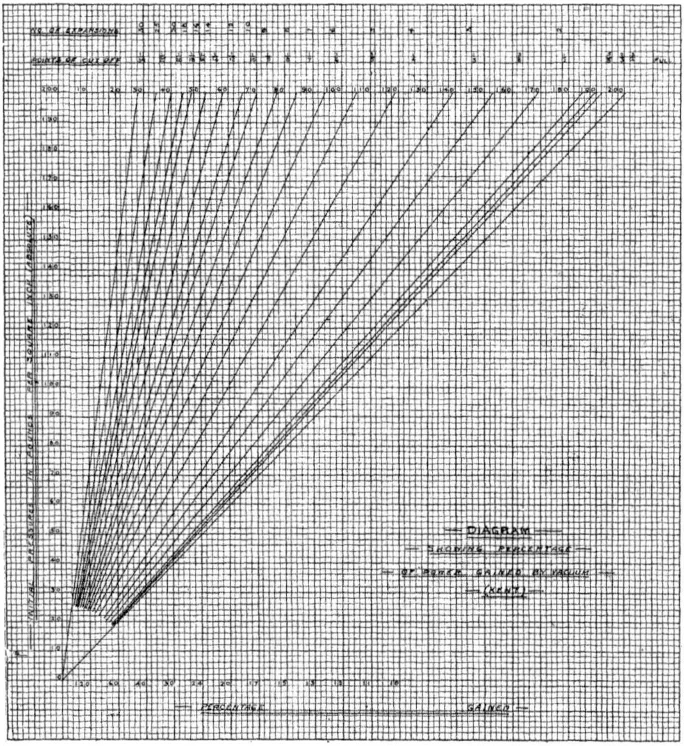


FIG. 58.

and often does, in large installations amount to the equivalent of one boiler, and where, as is often the case, machinery has been added until the boilers, in order to provide sufficient steam, are being forced beyond their economical evaporating limit, the addition of a condenser will ease them, and so effect economy in their life, as well as in their coal consumption.

**Jet Condenser.** — The most simple and theoretically most perfect form of condenser for steam is, of course, the jet or injector condenser, in which the exhaust steam is brought into direct contact with, and condensed by, water injected in such a manner as readily to absorb the heat.

The mixture of condensed steam and condensing water is then drawn away by the air-pump, and—if the water is sufficiently clear—a portion slightly in excess of the steam condensed is taken for boiler-feed, the remainder being usually run to waste.

The chief advantage of jet condensers lies in their relatively low first cost, and, as the amount of room they occupy is relatively small, they can, if fitted with an independent engine, be stowed in any convenient position. On the other hand, almost any water is suitable for condensing purposes, while it is necessary to be very careful what water is put into boilers, and, as in this country it is comparatively seldom that water suitable for boiler-feed can be obtained in sufficient quantities for condensing purposes, the adoption of the jet condenser renders it necessary to feed the boilers entirely with pure water, obtained either by an expensive filtering plant or at considerable cost from the public supply.

The amount of injection water required is found by a simple calculation, involving the sensible and latent heat of the steam at its terminal pressure on the one hand, against the initial temperature of the condensing water and the temperature required in the hot well on the other. In temperate zones the allowance usually made is from 22 to 27 times the weight of steam, and for the tropics 30 to 35 times or even more. The injection cock should always be made larger than is actually required, and should be graduated so that the engine-driver can regulate the amount injected until the best results are obtained. An engine working with a jet condenser should always be provided with a bye-pass to atmosphere, to be used when starting; by this means a vacuum may be formed in the condenser, and the "cooking" of the valves prevented by starting the injection water before the steam is turned on.

A common type of jet condenser is shown in section in fig. 59. The condenser is cylindrical, of cast-iron, having a circular opening in the top, corresponding to the bore of the exhaust pipe, for the admission of the exhaust steam, and a circular opening in the bottom for the escape of the condensed steam, condensing water and air to the air-pump. The injection cock is of brass, and the injection or spray pipe shown in section is of copper. It extends right across the condenser, and is pierced on its upper side with numerous ⅜-inch holes drilled so as to project the water upwards in numerous small streams to meet the incoming steam. The air-pump is single-acting with foot, bucket, and head valves. The liner is of brass, as are also the bucket and the valve seats. The bucket is packed by water grooves. The design and arrangement of the valves can be seen from the figure. This type of air-pump has been made without foot valves, but such an omission gives rise to a fluctuating vacuum.

**Ejector Condenser.**—The ejector condenser belongs to the same class as the jet condenser, in that the matter to be condensed and the condensing medium come into direct contact with each other.

There are several forms made, and they are exceedingly

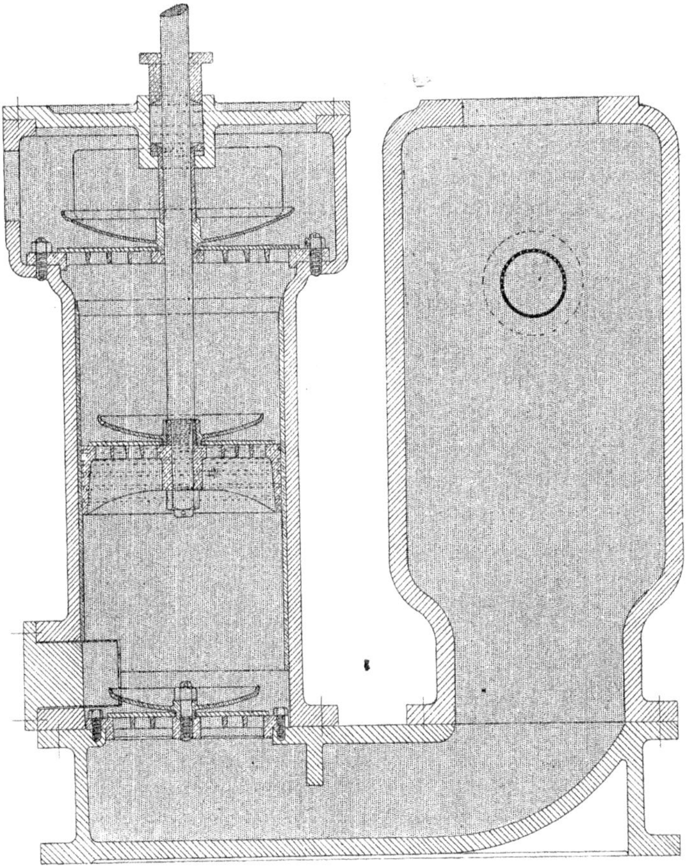

FIG. 59.

useful where circumstances militate against the employment of a surface condenser, or where first cost is the overruling consideration. Like the jet condenser, the quantity of condensing water required may be taken as a minimum to be twenty-two times the weight of boiler-feed. This, however, is the only similarity between the two, the ejector condenser differing from the other in the method by which the expulsion of the condensed steam, water, and air from the vacuum chamber is brought about. In the ejector condenser the momentum of the condensing water is employed to overcome the atmospheric pressure. It is introduced into the condenser in the shape of a solid jet, and is there surrounded by the steam, which is directed obliquely against it, and condenses on its surface.

There are three classes of ejector condensers:—overhead, self-acting, and adjustable capacity. In the first class the water is supplied to the condenser independently and under a head or pressure, and in the two second and more remarkable classes the necessary impetus is given to the water-jet by the exhaust steam itself. This is brought about by the pressure in the combining tube being lower than the pressure in the exhaust chamber surrounding it.

The overhead condenser is shown in fig. 60,[1] and is similar to the self-acting, with the exception that in the latter the water-inlet nozzles are made somewhat larger, and the holes in the combining tube are drilled at a sharper angle. The explanation is simple: the steam enters at the port *A* and surrounds the combining tube *B*, the water enters through the opening *C*, and, in the case of the overhead, rushes through the combining tube, and, drawing the steam through the small holes,

[1] Messrs Körting Bros., London.

condenses it and carries it and the air away through the nozzle *D*. In the case of the self-acting condenser, which will give a vacuum if simply immersed in water, the exhaust steam impinges on the water in the combining tube, and so gives it the requisite velocity to enable it to discharge against the atmospheric pressure. It is plain, therefore, that there is a minimum amount of steam below which it will cease to work, and great care must be taken in getting particulars as to steam consumption, etc., previous to installing it.

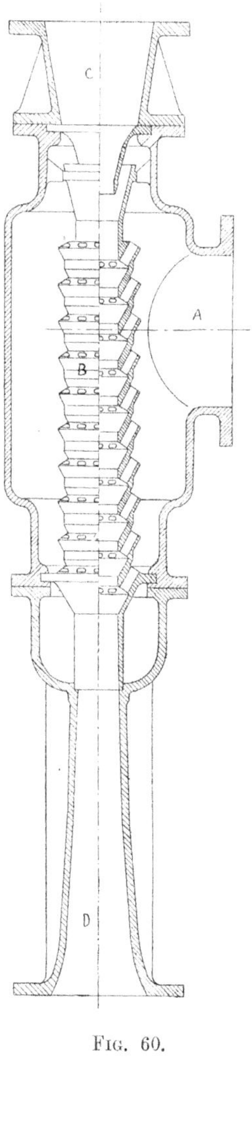


FIG. 60.

Where the load on the engine, and consequently the consumption of steam, varies suddenly and at irregular intervals, the overhead condenser should be used, as it cannot fail to give a good vacuum provided it is suited to the conditions of maximum load. Where the load is constant, and there is no natural head of water, the self-acting condenser is the better. Until recent years these were the only two types in use, and the ejector condenser was limited to special cases and special engines. In 1891, however, Messrs Körting Bros. invented their adjustable capacity condenser. It is illustrated in fig. 61. The means of adjusting the condenser to its load consists of the sliding nozzle *G*, which can be moved up and down the combining tube by the handle *H*, and so cover or uncover the successive rings of holes, and decrease or increase the capacity of the condenser at will. The nozzle can be locked in any position by the key *I*. The reference figures are:—*A* Steam Inlet, *B* Combining Tube, *C* Water Inlet, *D* Water Outlet, *F* Steam Jet for starting purposes when the water has to be lifted.

With the overhead condenser a vacuum of from 27 to 28 inches can be obtained, and with the self-acting and adjustable capacity 25 to 26½ inches is good.

**Surface Condenser.**—A section of the ordinary type of surface condenser is shown in fig. 62. It is circular in section, and the exhaust steam passes into the condenser through the circular opening *A*, where, striking the baffle plate *B*, it spreads itself through the condenser, and by contact with the tubes through which the condensing water is circulating, becomes condensed, and is drawn away by the air-pump through the circular opening *D*.

The condensing water enters the condenser at the bottom through the opening *E*, and encountering the baffle plate *F*, passes through the tubes to the opposite end of the condenser, whence it returns through the upper half of the tubes, and escapes through the opening *G* at the top.

By having the coldest tubes at the bottom, a lower temperature and better vacuum can be obtained than otherwise. In small condensers the condensing water usually enters at one end, at the bottom, and passes out of the other at the top, only making one passage through the tubes.

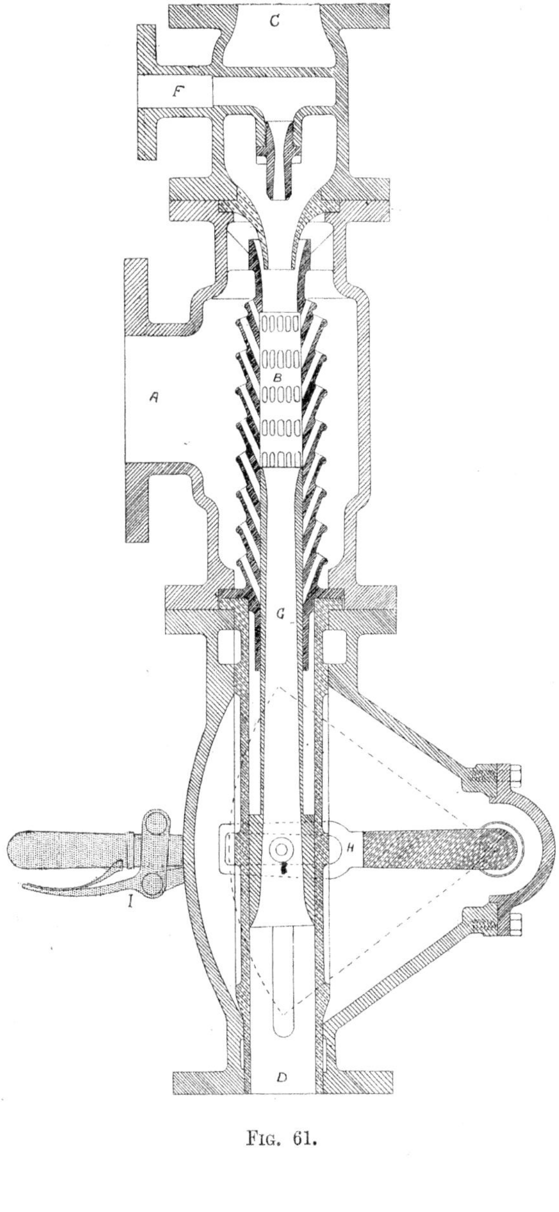


Fig. 61.

Handholes for examining the tubes and cleaning the condensers are shown at *H*. The bye-pass valve *K* is provided in some cases in order to pass any make-up feed-water required into the steam space. The tubes, when coated with fat, oil, or grease from the cylinder, can be cleaned by throwing the condenser out of work, and, after filling the steam space with water, introducing a quantity of caustic soda. If the solution is warmed up by a steam jet, the coating will readily come off the tubes, and the water can be run out through the mud doors. Some-

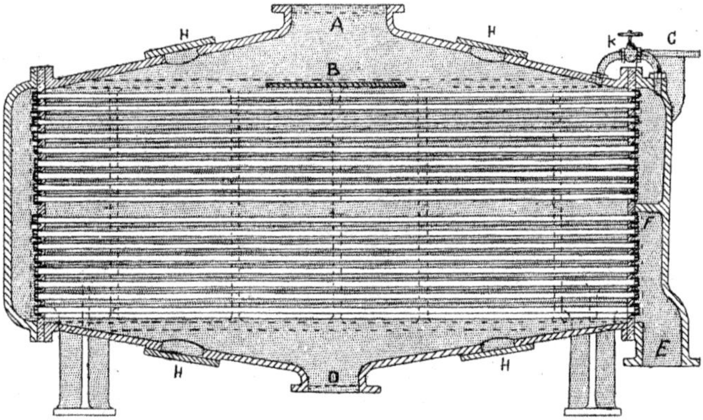


FIG. 62.

times the caustic soda is introduced while running, but this means is not so effectual.

**Evaporative Condenser.**—The principle involved in evaporative condensers is that of allowing the steam to be condensed to expend its sensible and latent heat in heating and evaporating at atmospheric pressure a small quantity of water, the evaporation being assisted by a current of air driven at a high velocity over the surface of the pipes containing the steam. Both the air and the

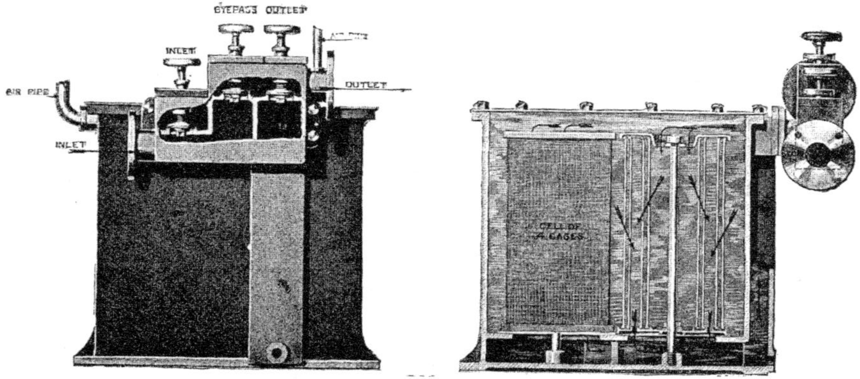


FIG. 63.

To face page 183.

water are passed over the outside of the tubes in which the steam is condensed.

The advantage of this type of condenser lies in the fact that, with a surface only slightly in excess of that in the surface condenser, the weight of condensing water consumed only slightly exceeds, under the least favourable conditions, the weight of steam condensed. In order that the natural breeze may assist the evaporation, the open type of evaporative condenser is usually placed on the roof of a building. With the employment of a fan this plan becomes no longer necessary, and the condensers may be placed in any convenient position.

**Feed-water Filters.**—Condensed steam always contains oil which has been used for the lubrication of the cylinders. If this oil gets into the boilers, it will cause corrosion and great trouble, and it should be removed before the feed-water enters the boilers. To effect this, several filters have been designed. Many of these are placed between the feed-pump and the boiler, and are worked at the same pressure as the boiler. This, however, is objectionable, because, as the filtering medium collects the oil, the passage of the water is obstructed and the pressure of the feed-water carries much of the impurity through the filter.

Fig. 63 [1] illustrates Messrs List & Munns' filter, which is of the non-pressure type, and which has given good results in practice. It acts by gravitation alone, and the impurities in the feed-water have time to settle on the filtering medium. As the filtering surface is very great, a large amount of deposit does not materially obstruct or diminish the speed of flow through it. The illustration shows the filter in section, and the arrows

[1] Messrs J. H. Carruthers & Company, Glasgow.

indicate the course of the water through the filter. This filter is preferably placed at a lower level than the hot well into which the air-pump discharges, and in the feed-circuit between the hot well and the feed-pump, never between the feed-pump and the boiler. These filters can be arranged either singly, in duplicate, or in series, and they are usually fitted with valves so arranged that one or more of the filters can be cut off the feed-circuit for cleaning purposes while the others continue working. The filtering medium is easily cleaned and is cheap for renewal.

## THE RECOOLING OF CONDENSING WATER.

Users of refrigerating machinery are necessarily aware of the advantage of an ample supply of water for the ammonia and $CO_2$ condensers; this is, however, not always available, and numerous attempts have from time to time been made to economize the water used for condensing purposes by recooling it and using it over and over again. The question is one that is of considerable importance to owners of refrigerating plant, and is therefore dealt with at some length.

**Worthington Cooling Tower.**[1]—In this tower the water to be cooled is discharged in a heated condition to the top of the tower, where it passes through a distributor of the sparger type. By this it is scattered in fine drops resembling rain, and trickles down through the cellular tower, which is usually made of galvanized iron pipes 4½ inches in diameter. In some instances glazed earthenware pipes 6 inches in diameter are used. These pipes are stacked in tiers breaking joint one with the other. The whole cooling surface is enclosed in an iron or steel

[1] Messrs the Worthington Pumping Engine Company, London.

tower, and mounted over a tank, into which the cooled water runs. The cooling of the water passing through the tower is due to the radiation from the pipes and the sides of the tower, the contact of the cool air supplied by a fan with the heated water, and the latent heat absorbed in evaporating a portion of the water. The last named is, of course, by far the most potent factor. The following figures relate to the performance of one of these towers erected for the Huse & Loomis Ice and Transport Company, St Louis, Mo., U.S.A. They are the averages of observations taken every day for a period of fifty-four days. The readings in the original were in U.S. gallons of 8⅓ lbs., but they have, for simplicity's sake, been converted into Imperial gallons. The trials were taken during August and September. The temperatures are in degrees Fahrenheit.

| | |
|---|---|
| Temperature of air, | 78·2 |
| Temperature of city water, | 81·5 |
| Temperature off condenser and in tower, | 119·3 |
| Temperature on condenser and out of tower, | 76·3 |
| Water cooled below temperature of air, | 2·1 |
| Reduction of temperature of water, | 43·2 |
| Gallons circulating water per minute, | 375·5 |
| Gallons city water used per 24 hours, | 88,414·9 |
| B.T.U. carried off per minute, | 161,352 |
| Back pressure ammonia, | 15·1 |
| Forward pressure ammonia, | 181·1 |

Messrs Worthington also make a natural draft tower, the general principles of the design being similar to the forced draft tower.

**The Barnard Cooling Tower.**[1]—The outer casing, or

[1] Messrs the Wheeler Condenser Company, London.

tower proper, is a rectangular structure in steel, or iron plate, wood, or brickwork. Inside this structure is suspended vertically, and at regular intervals, a series of steel-wove mats, galvanized after weaving. The water to be cooled is pumped to the top of the tower, and distributed over the mats, down which it trickles, flowing over and round the wires, and falling into the tank at the bottom. A fan propels a body of air up between the mats, and the water is cooled in the manner already mentioned.

In a test recently carried out with one of these towers, the shade temperature of the atmosphere was from 80 to 90° F. The water was delivered to the tower at a temperature of 120° F., and was cooled down to 78° F. The fans ran at 225 revolutions per minute, absorbing power to the extent of 1¼ per cent. of that of the main engine. The vacuum maintained in the condenser was 26½ inches.

**The Klein Cooling Tower.**[1]—The Klein coolers are of three leading types, *viz.*:—the open or self-ventilating type, the natural draught type, and the fan-draught type. In the open type the water is pumped to a height of about 20 feet, in the natural draught type to 25 feet, and in the fan-draught type to from 30 to 35 feet. The open type is the cheapest, but occupies the largest ground space, the area required being 2 square feet per I.H.P. The natural draught tower is the type which is generally found most useful, and occupies about 1 square foot per I.H.P., the total height of tower being from 65 to 70 feet. The fan-draught tower is used in circumstances where space is very restricted, and occupies about ·25 to ·5 square feet per I.H.P., height being about 45 to 50 feet.

[1] Messrs the Klein Engineering Co., Ltd., Manchester.

In all three systems of cooling the system adopted is to first pump the water into one main trough, from which it is divided into two, three, or more distributing troughs; from these it passes to a large number of distributing gutters. From these gutters the water falls in streams on to frames consisting of narrow wood laths of suitable width and placed at suitable pitch vertically, the pitch of the frames in a horizontal direction being about 3 inches to 3½ inches.

The amount of water lost by evaporation varies according to the type of cooler employed, and is generally found to be less with the natural draught tower. It does not in any case exceed 3 per cent. of the water in circulation, and is often considerably less.

**Körting Coolers.**[1]—This apparatus consists of a series of small nozzles constructed as shown in fig. 64. The water to be cooled is forced by a pump under a pressure of from

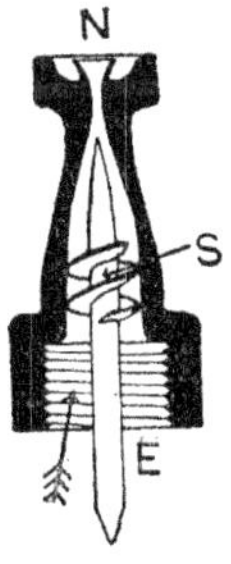


FIG. 64.

7·5 to 10 lbs. per square inch through the nozzles. The spindle *E* being fixed, the water travelling round the spiral *S* acquires a rotary motion, and, on issuing from the nozzle

[1] Messrs Körting Bros., London.

*N*, centrifugal force causes it to fly into small drops, thus exposing a large surface to the air. A small portion of water is evaporated from each drop, and the remainder is cooled by the latent heat thus absorbed.

It will be seen that the efficiency of all these apparatus depends on the humidity or aridity of the atmosphere. If the latter conditions prevail, the water issuing at any temperature from the nozzle will be cooled down to the temperature of the air, but if the air is already saturated there will be less absorption of moisture, and the cooling will not be so efficient. Nevertheless, the temperature of the water will always be reduced to within a few degrees of the surrounding atmosphere, and even in the worst cases the difference of temperature between the cooled water and the atmosphere has been found not to exceed 10° F. The loss by evaporation is from 3 to 4 per cent. of the water cooled.

From a theoretical point of view it would be desirable to divide the hot water into a very fine spray, and this object can be very easily attained if the velocity with which the water is forced through the nozzles is increased. But there are practical objections:—first, a fine spray is easily carried away by the wind, and further, as the cost of pumping increases in proportion to the pressure obtained, it is desirable in practice to reduce the pressure to as low a figure as possible. The minimum pressure at which good efficiency can be obtained is 7·5 lbs. per square inch. A point of importance with regard to these nozzles is that, as the air must be allowed free access to every spray, it becomes necessary to distribute the nozzles over as large an area as possible, and experience has shown that under ordinary conditions a minimum of 50 square feet of area is necessary for every 1000 gallons cooled. Where circumstances permit, these nozzles are generally arranged over a tank on the roof, so that the water when cooled falls into the tank, and thence to the condenser.

The exact arrangement by which the re-cooled condensing water coming from the cooling tower is made use of is obviously a matter of convenience. Sometimes separate towers are used—one for the ammonia or $CO_2$ condensers, and the other for the steam condensers; but the more common way is to run the cooled water first over the condensers of the refrigerating agent and then through the steam condensers, from which it returns to the cooling tower.

## PUMPS.

Theoretically a pump will draw water from a depth of 34 feet, but, owing to the imperfect vacuum caused by valve leakage, the maximum suction lift obtainable in practice is from 20 to 24 feet. In pumping hot water the pump should always be given a suction head—*i.e.*, the water should flow into it by gravity. Foot valves should always be provided on the suction of any pump where the lift exceeds, say, 8 or 10 feet, and arrangements should be made so that the suction pipe may be charged before the pump is started.

Small direct-acting pumps are very inefficient and consume a vast quantity of steam. A piston speed of 100 feet per minute is common in practice, but this is high for small pumps with short strokes. For large pumps with ample valve area it may be considerably exceeded.

**Boiler Feed-Pumps.**—In deciding upon the proper size of pump to instal for boiler-feed purposes, the amount of water the boiler is capable of evaporating, not the horse-power of the engine, should be taken as a basis, and the pump should be large enough to comfortably supply this.

Duplex direct-acting pumps are frequently used for the boiler-feed, and though they use a large amount of steam they work well.

Fig. 65 [1] illustrates the Weir boiler feed-pump, which has recently come largely into use. Owing to a positive valve gear the pump will start working immediately steam is turned on. The pump runs at the slow speed of 12 double strokes per minute, and the valve gear ensures full length of stroke and constant action. Steam is used expansively, and as the piston automatically slows down towards the end of the stroke, there is no shock or jar, and the valves are enabled to settle quietly on their seats.

**Air Lift-Pumps.**—A deep well-pump which has come into considerable use during the last few years, and which does not suffer from the deficiencies which have made the deep well bucket-pump so notorious, is the air lift-pump. There are several of these in the market, but, as the principle is the same, a description of one will be sufficient. Fig. 66 [2] illustrates the Bacon air lift-pump. It consists of two plain pipes which are let down into the well. The pipes are connected at the bottom into a bell-mouthed piece as shown in the illustration. The outer pipe contains the air, and the inner one is the delivery pipe for the water. There are no valves of any description except, of course, in the steam-engine and air compressor, and as no standing water remains in the pipe when not at work, there is no danger from frost. When working, air, in excess of the pressure due to the head of water in the well, is forced down the outer pipe into the mouthpiece; from this point the air rises in the delivery pipe, and, forming air pistons, carries the water with it to the surface.

[1] Messrs G. & J. Weir, Cathcart, Glasgow.

[2] Messrs Blake & Knowles' Steam Pump Works Company, Ltd., London.

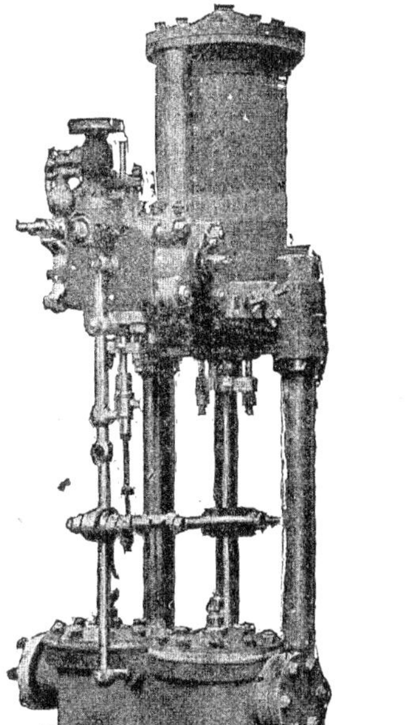

Fig. 65. *To face page* 190.

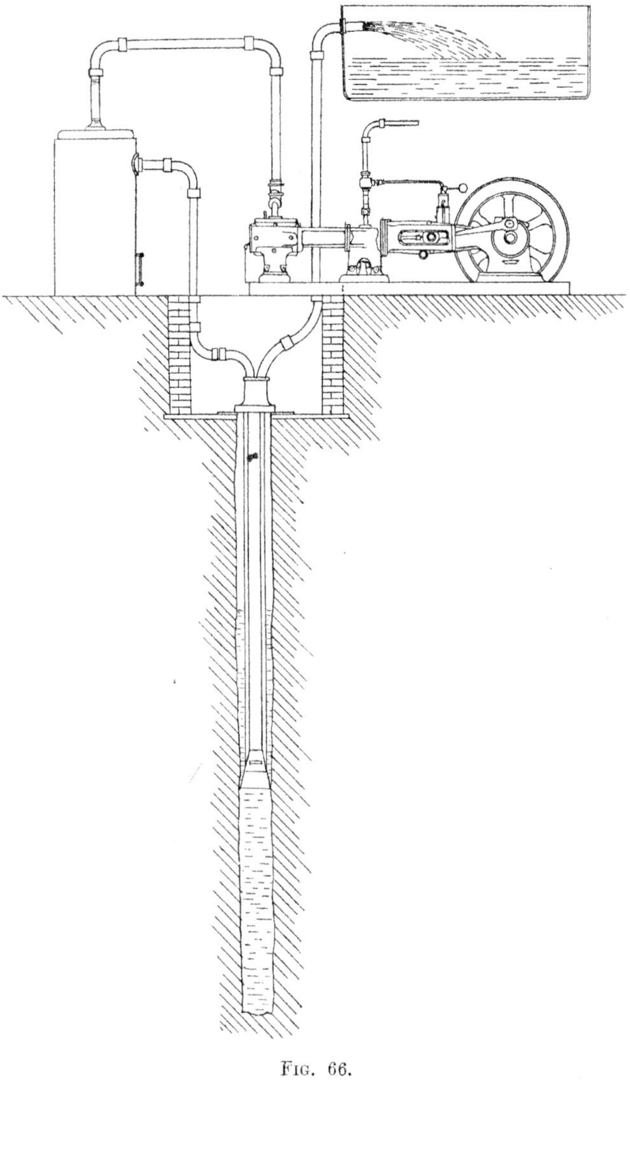

Fig. 66.

The pressure of the column of air and water in the delivery pipe being less than that of the solid column of water outside, the water flows continuously to the surface. As the air pistons rise in the delivery pipe, they expand in direct proportion as the load upon them becomes lighter, and, when discharged, they are at atmospheric pressure. A simple illustration of the action of these air bubbles can be seen in any soda-water syphon just as it empties. The gas forms air pistons in the central glass tube, and they carry up the soda water with them, expanding as they rise. The expansion of the air when doing work has a remarkable cooling effect on the water raised, a result which is all gain when the water is used for condensing purposes. The air compressor may be situated at any distance from the well, and is, as a matter of fact, usually placed in the engine-room with the other machinery.

Before installing these pumps it is necessary to obtain the fullest possible information of any capacity test to which the well has been subjected, and especially the difference in the levels of the water which have occurred while any particular amount of water is being pumped from it.

Those methods of calculation which have given the best results in practice are as follows. The first thing to do is to ascertain the level at which the water will stand in the well while pumping is in progress. Supposing, for instance, a well has a static water-level of 60 feet, and when, say, 5000 gallons per hour are being pumped, this level drops to 100 feet from the surface; under these circumstances it is necessary, in order to obtain sufficient submersion of piping, to double the head of 100 and add from 10 to 20 feet. The bell-mouthed uptake pipe would then be 210 to 220 feet from the surface. This will require an air pressure of about 100 lbs. on the square inch, and will secure a continuous and satisfactory flow of water. It is usual, on account of the head necessary before the pump will work, to only deliver the water to the surface, any additional pumping that is required being done by another pump of the ordinary type; but if it is essential, for any reason, that the pump should deliver to a point above the surface of the well, this extra head must be included in the head to be pumped against and doubled accordingly. For lifting the water against a head of about 50 feet, 2 cubic feet of free air per minute is necessary for each cubic foot of water raised; for a head of 100 feet, 3 cubic feet; for a head of 150 feet, 4½ cubic feet; and for a head of 200 feet, 6 to 7 cubic feet is allowed. Assuming, therefore, that 5000 gallons per hour are being raised 100 feet, an air compressor delivering 40 cubic feet of air per minute would be necessary, or, as it is always desirable to allow a margin, a slightly larger size than this should be installed. As the air, even when it is delivered to the well just as it comes hot from the compressor, cools the water some 10 to 12° below its initial temperature, it is probable that if the air were first cooled down to atmospheric temperature by artificial means, the water in the well would be cooled considerably lower still. The size of pipe selected is one that will give the least possible friction to the flow of water passing through it. The capacity of the pump is practically unlimited, and with a properly proportioned plant it is said to work efficiently up to sizes of 30 inches diameter; considerable power is, however, required.

## Electric Lighting.

It is not too much to say that, had it not been for the electric light, the huge cold stores as at present designed would not have been possible. Light with an absence (comparative) of heat and combustion has enabled daylight to be entirely dispensed with, and now cold stores are so designed that not a ray of daylight ever penetrates them if it can be prevented.

The use of electricity, however, is not confined to lighting, and it is becoming more and more the practice to drive the subsidiary machinery, such as lifts, hoists, fans, etc., by electric motors instead of by hydraulic power or by belting from the main engine. The chief charm of electricity is its flexibility, and there is no occasion to study lining out as with shafting, or to take any notice of gravity or frost.

The generating set should preferably consist of an engine and dynamo mounted on the same bed-plate and coupled directly together. This design occupies the smallest floor space for a given power. The combined set will run at a speed, depending on the type of engine, of from 200 to 450 revolutions per minute, and should give an efficiency of 80 per cent. The dynamo may be either shunt or compound wound; the former is satisfactory if the current is only wanted for lighting and the load is likely to be regular. If, however, there are motors on the circuit, and the load is liable to fluctuate, it should be compound wound so as to secure the best self-regulating properties and maintain a constant voltage. Fig. 67 illustrates a handy little combined set for electric lighting and power. The speed is 400 revolutions per minute, and the voltage 110; the dynamo is compound wound; the engine is of 25 B.H.P., and the total floor space occupied is only 6 feet 6 inches by 3 feet. The voltage may be either 100, 110, 200, or 220; but if the area to be lighted is not great, 110 will be found very suitable, and the lamps are

Fig. 67.

*To face page* 194.

more efficient and last longer than those of a higher voltage. Further, if the lamps have to be put in bulkhead fittings placed flat or at an angle, as is common in cold storage work, the stronger filament of the 110-volt lamp is not so liable to sag and break off as those of the higher voltage lamps have been found to do. There is nothing about the switching arrangements for cold storage work that requires special mention.

The point of most importance in cold storage work is the wiring. This should be exclusively done with high conductivity tinned cables insulated with india-rubber, as no other insulation stands the work so well. The mains should be run either in American white wood casing or in waterproof iron piping or steel conduit. These latter have some advantages, but iron and rubber do not do well together, and for ordinary cold storage work the former has been found quite satisfactory, and is certainly much cheaper and easier to put up and adapt, or alter should occasion necessitate it.

The lighting inside the cold stores should be with incandescent lamps exclusively, the mains being run as much as possible in the corridors. The yards, engine-rooms, etc. may be lighted with arc lamps, those of the enclosed type being best. On a 110-volt circuit the enclosed arc lamps should be run in parallel on open circuit across the mains, as they are most efficient at this pressure.

In the majority of cases, however, the incandescent lamp will be found to give the softest and most comfortable light.

Ship or oyster fittings are most usually used, protection being necessary to prevent the lamp being broken when goods are being moved about; but the number, type, and position of the lights and fittings are questions that cannot be generalized and can only be decided for specific cases.

There is one point, however, that should receive more attention than it has, and that is the practice of keeping the lights burning in cold stores whether the men are working in them or not. In some stores it is the practice for some reason or other to keep the lights always burning. This means not only a consumption of steam in the electric light engines, waste of lamps, etc., but also a consumption of steam in the refrigerating machines. It has been found by experiment[1] that a 16-c.p. lamp will give off heat equal to about $\frac{1}{60}$ ton refrigerating, while the lights in a well-lighted room will give off about $\frac{1}{12}$ of the total refrigeration required for the room.

This would be bad enough if there were any necessity for it, but there is not, and there is no reason why each room or floor should not have its own master-switch placed in a convenient position so that the lights could be switched off when the rooms were not being worked.

## Electric Motors.

The controlling feature of an electric motor is the way its magnets are wound. This depends on the kind of work it has to do.

Motors may be either shunt-, series-, or compound-wound.

**Shunt-wound Motors.**—This type of winding is the most common, and is applied to motors that have to run at a constant speed independent of the load. The speed of the motor will only vary some 3 to 5 per cent. between

[1] Stoddard on "Insulation," *Proceedings American Warehousemen's Association*, 1901–02.

no load and full load; it is the most simple kind of winding, and has the most simple regulating devices.

**Series-wound Motors.**—Series-wound motors are most useful for heavy work where a heavy starting effort or torque together with a large variation of speed, or where reversibility, is required, as, for instance, in cranes.

As the current passing round the field-magnets is the same as that passing through the armature and is dependent on the amount of work the motor is doing, the speed will vary with the load, and the motor will slow down when heavy loads are thrown upon it, and run up when the load is light. This characteristic of decreasing in speed with a heavy load tends to keep the load fairly constant and at the same time protects the motor, as, owing to the drop in speed, the horse-power given out will not increase under full load in anything like the same proportion as it would in a shunt motor where the speed is constant.

**Compound-wound Motors.**—These motors combine the advantages of the two former, and are most useful where they have to drive machinery which has a fluctuating load. The combination of the two windings prevents heavy rushes of currents when an extra lot of work comes on, and at the same time keeps the speed constant within reasonable limits.

They are much used for driving machine tools that have to be stopped and started under load, or reversed. Though the controlling arrangements are rather more complex than those for the shunt-wound motor, they are most suitable for general purposes. They can be compounded to any extent, so as to fit any special purpose.

For cold storage and other work motors of the enclosed or protected type are much the most suitable. They are more compact, and, while being very fully protected from mechanical injury, are fully ventilated. Sometimes completely enclosed motors are put in, but these are not necessary except under very special circumstances.

## Electric Lifts.

Electric lifts have largely come into use in place of the hydraulic lifts. They are safe, reliable, capable of easy control, and noiseless, and they compare favourably in first cost with the hydraulic elevator.

The principal advantages, however, are ease of fixing and economy in working.

A hydraulic cylinder requires careful fixing, and in order that the ram may be sufficiently long and the shieves sufficient in number, the travel must be predetermined.

With an electric lift the travel does not affect the gear, and increased travel merely means a longer rope.

With a hydraulic elevator power water has to be used in the cylinder whether the load is light or heavy. This has been avoided in special cases in some designs by having a small ram for lifting or lowering empty, but it is not usual.

With electric elevators the current used is in direct proportion to the load, and as the lifts are nearly always fixed on the overbalance principle—that is to say, the car is balanced—when the average load is in the car the lift has no gravity work to perform, and the motor is only exerting its full power when the car is travelling down empty or has its full load on the up journey—conditions which do not always exist in practice. A rule frequently followed is for the balance weight to equal the weight of the empty car plus one-third to one-half of the live load. This, of course, can be varied to suit the circumstance of each case. In all cases the object arrived at is to so balance the car that for its average load the motor will not be doing more than one-sixth or one-eighth of its full power.

The electric motor and gear can be placed either at the bottom or top of the hoistway; in the majority of circumstances the latter is preferable. In this case one face of the winding drum is placed plumb over the centre of the hoistway, and the other over the counterbalance weight. It is the usual practice to have the balance weight and car ropes wound on the same drum, the one winding up from one end, while the other unwinds, and *vice versâ*.

The power is usually transmitted from the motor to the drum by a phosphor bronze worm and wheel. This is not an efficient mechanical device, but it is safe and has proved the best in practice. The worm shaft should have collar bearings to take the thrust, and all bearings throughout should be made with caps so that any part can be easily removed. This, however, is a maker's detail.

The control of an electric elevator is simplicity itself. A magnetic brake is generally fitted; this is always in gear, and is only released when the current is switched on and the motor is at work. Shunt-wound motors are usually used. Sometimes a few series windings are provided so as to increase the torque at starting, these being cut out when full speed is reached.

The method of starting and stopping the elevator varies with the class of work for which it is required. Goods elevators, such as used in cold stores, are usually operated by a rope passing through the car. This actuates a starting switch by means of which the resistance in series with the armature is cut out step by step. In other designs, and especially when passengers are carried, the starting switch is placed in the car and worked by the attendant, who switches out the resistance slowly. The safety devices for an electric lift are generally similar to those used in steam or hydraulic lifts, and consist usually of some form of cam gear, connected to two rocking shafts, to which the hoisting ropes are fixed. If the ropes break or become slack for any reason, these rocking shafts are pulled in by a powerful spring, and the cams on their ends engage with the steel girders or bury their teeth in the woodwork backing supporting them. Tests of this safety gear have shown that when the hoisting ropes are suddenly disconnected, the car only falls half an inch before being brought to rest. To prevent the car over-running itself stops are fixed on the starting rope at the limits of the travel, but, as an additional precaution, knock-out switches are placed just above and below the limits, so that, if the car by any chance passes the stop on the rope, it knocks out these switches and interrupts the current to the motor; this, of course, stops the car.

Sometimes accidents have occurred from working-men falling down the lift-wells, and one of the advantages of the electric elevator is the ease with which safety doors may be fixed. These are arranged so that the opening of any door breaks the main circuit, and it is impossible for a door or gate to be opened unless the car is in position opposite it. It is equally impossible for the motor to be started and the lift moved unless the gate is locked and the circuit remade.

# CHAPTER VIII.

## INSULATION.

**The Objects of Insulation.**—Space, enclosed by walls which have been well insulated, can be readily reduced in temperature and easily kept cool. The more carefully and effectually the heat is excluded, the smaller will be the demands on the coal supply, and the greater will be the returns on the capital invested. For this reason the insulation, like the machinery, should be looked upon as a permanent investment. The expenditure is made once and for all, while the flow of heat, which, in the absence of good insulation, will take place, is continuous and unceasing. It is not desired or intended to recommend that first cost should be lost sight of, or that *elaborate* precautions should be taken to keep out the heat. A study of nature and her laws will show that, as a rule, the more simple the expedients adopted, the more likely they are to be successful. Frequently large sums of money have been expended in some costly and complicated arrangement of insulation, the result of which is not nearly so efficient as a more simple and cheaper arrangement of the same materials would have been.

The whole question is one of capital account *versus* revenue account. If the insulation is scanty, the sum on which dividends have to be paid is small, but, in order to keep the necessary temperatures, so many extra tons of refrigeration, and consequently so many extra tons of coal, besides oil, stores, and labour, are required each year and reduce the amount available for dividends. If the insulation is good, the capital is greater, but the working charges are less, by the value of the coal, etc. saved. The problem to be considered, therefore, is how much money can be saved per annum on revenue account by spending a definite extra sum of money on capital account. This is probably the most important point which comes under the engineer's consideration, and is one which, with our present knowledge of the heat-resisting properties of insulation, can only be settled in a satisfactory manner by an actual experience of the results which various designs have given.

The flow of heat through any substance is proportional to the difference in degrees Fahr. between the temperature on the two sides of that substance, or, in other words, to the temperature head. There is no substance known which will absolutely stop heat, and, therefore, the matter becomes one of degree.

The insulation of a cold store consists of the shell of the building, the filling material, the timber necessary to hold it in place, and the paper necessary to keep it dry and air-tight. Each portion is as worthy of study and consideration as another.

**Various Insulators and Filling Materials.**—The value as an insulator, of any particular filling material varies to a great degree with its specific gravity, and, generally speaking, the lighter the material, the better insulator it is. This rule holds good until a point has been reached where the material is either so light in itself or is so loosely packed that it allows the air to circulate, and so nearly conforms to the conditions which prevail in an open air space. A specific gravity of 0·160 is stated by Mr Starr[1] to be the limit below which it is not advisable to go.

For preventing conduction and convection of heat from one surface to another, still air is the best material that nature has placed at the engineer's disposal. This statement does not hold good as regards radiant heat, but, as that seldom comes into account and can be guarded against in other ways, it is not very material. Unfortunately, air if left to itself will not remain still, but will circulate in currents, and so carry heat by convection. It is consequently necessary, in order to make use of the insulating properties of air, to find some means of preventing this circulation, and the best insulating substance therefore is that which for a given weight and volume will contain the largest quantity of air in the most minutely subdivided particles. There are several substances which fulfil this condition to a greater or less degree of perfection, and they have been adopted as insulators or fillers in connection with refrigerating work. The most prominent of these, given in alphabetical order, are charcoal, granulated cork, sawdust and shavings from hard and soft wood, and silicate cotton or slag wool. Other materials, such as hair-felt, cow hair, rice husks, silicated straw boards, calcined pumice-stone, red fibre, sanitary mattress (curled and dyed) have been experimented with, but as they are not in general use it is hardly necessary to refer to them in detail. Sawdust is familiar; the principal characteristics of the others are as follows:—

**Charcoal.**—Wood charcoal is an impure form of amorphous carbon, and, in addition to a small quantity of

[1] *Proceedings American Warehousemen's Association*, 1901–02.

oxygen and hydrogen, contains a considerable portion of mineral matter. It is obtained by heating wood in the absence of air. There are in general use two kinds of charcoal—that made from shavings, turnings, etc., and that made from the carbonization of small pieces of wood not specially prepared.

Charcoal, while being inodorous in itself, possesses the power of absorbing gases and odours to a considerable degree, which makes it exceedingly valuable as a filtering material. It is exceedingly porous, and it contains a great quantity of air in minutely subdivided particles. It is to this property that it owes its high position as an insulator. Flake charcoal, which is largely employed for this purpose, is manufactured from wood turnings, birchwood being principally used. The shavings are carbonized in mechanically fed retorts, and the charcoal is afterwards cooled by a special process which ensures that it will take up sufficient oxygen to avoid its subsequent reheating.

When sent out by the manufacturer,[1] the best quality flake charcoal weighs about 10 lbs. to the cubic foot, and is packed into the insulation to a density of from 12 to 13 lbs. per cubic foot. If subject to vibration, charcoal will settle down to a still greater density. This settlement can be remedied in a simple manner by removing the top boarding of the insulation and filling in the vacant space with new material. When first put in place, the charcoal is filled in between the timber casings, and is consolidated either by being rammed down with sharp-pointed sticks or by vibration caused by hammering the containing woodwork. The former method results in tighter packing, but is apt to break up the flakes and cause powdering. By the latter method the charcoal is

[1] Messrs the Cartvale Chemical Co., Ltd., Paisley, N.B.

not packed quite so tightly, but it escapes damage and gives at least as good an insulation with a smaller weight of material. Charcoal has no effect on metal, and can be packed directly against a ship's side or a steel bulkhead or girder without damage either to itself or the metal.

**Cork.**—Cork is produced from the bark of the cork tree, a variety of the oak family indigenous to, and largely cultivated in, Spain, Portugal, and other countries of the Mediterranean region. The peculiar formation of the bark is due to a natural process which, when expressed in botanical language, appears too alarming to be stated here; the result, however, is the formation of successive layers of cork cells which are filled with air. The tree—which lives to a great age—is first stripped of its bark at the age of from 10 to 15 years; this is virgin cork, and is useless for anything but what are alleged to be ornamental purposes. Every subsequent seven years the tree is again stripped, the curved strips are flattened by weights when under water and are charred or steamed to close the pores. For this latter purpose the bark is placed in large copper boilers.

Granulated cork, with which refrigerating people are more intimately concerned, is a bye-product, and is made from the shavings and chips produced in the making of corks, as well as from the commoner qualities of cork wood. To produce it, the chips, etc. are ground in ordinary grinding mills, care being taken to free the fibre from the bark, which is very gritty, and from the belly. It is then sifted or screened to separate the various sizes of grain, the coarsest and medium sizes being most commonly used for insulating purposes.

As an insulator cork is very valuable. It is exceedingly light, and it packs naturally to about 5 lbs. to the cubic foot. In order, however, to prevent silting, it is desirable to pack it to a density of about 6 lbs. to the cubic foot.

**Silicate Cotton.**—Under the various names of rock cotton, slag wool, granite rock wool, etc., silicate cotton has been known for years, and has, by merit, gradually forced its way to the front as an insulating material. It is a mineral fibre of the nature of spun glass, made from the slag of the blast furnaces. Granite rock cotton is made from crushed granite mixed with a small proportion of limestone.

The process of manufacture of silicate cotton was, it is believed, discovered by John Player, of Stockton-on-Tees, in the later sixties, and in 1870 he patented a process for the manufacture of the material. Silicate cotton should not contain much lime or it will become brittle and liable to pulverize, and for this reason that made from a furnace slag, which is moderately free from lime, is the better. In the process of manufacture molten slag or granite and lime is allowed to flow from the cupola through an orifice, where it comes in contact with a high-pressure steam jet which disintegrates it and blows it into a soft mass of fleecy fibres, which are mixed up and intertangled with each other, thus imprisoning a large amount of air in and between the interlaced fibres. The temperature of manufacture is about 3000° F., and 1 cubic foot of rock or slag will make about 12 cubic feet of the finished material. As it is entirely a mineral product, it is not in any way liable to be attacked by, or attractive to, vermin, and it will not harbour animal life. Its insulating properties are considerably damaged if it gets wet, but as this applies to all other filling materials, and as, if paper is properly used in erecting the insulation, dampness will not occur, it is a point to which no material comparative importance need be attached. A great recommendation to the use of silicate cotton is the fact that it is absolutely fireproof, and—and this is most important—it is entirely without smell and will not absorb odours. Considerable complaint has at times been made that silicate cotton, when used as a filling material, will settle in the insulation, and leave blank air spaces through which heat can pass both by conduction and radiation. There has probably been some foundation for this charge, but it has probably arisen from the workmen having carelessly packed the wool too loosely. If packed to its proper density, the tendency to settle is not great, especially when it is remembered that the slag wool is in itself light, and that it has the skin friction of the studs and the matchboarding on both sides to hold it up. Even if settlement does take place, after a few years the top of the insulation can be easily opened, and the space, left empty by settlement, filled with new wool.

**Principles to be Observed.**—The qualities of good insulation may be enumerated as follows:—

1. It should have a high specific heat.
2. It should be light in weight, free from unpleasant odours in itself, and a non-absorbent of external odours.
3. It should be non-hygroscopic, and should not hold moisture in suspension.
4. It should have no attraction for vermin.
5. It should not be liable to spontaneous combustion or contain acids liable to attack metals.
6. It should not be liable to decay or to settle after it is put in place.
7. It should consist of various laminæ, so as not to present an obstruction of uniform density.

8. It should present an air-tight surface on the side exposed to the high temperature or source of heat, and an insulated layer on the inside.

Though these requirements at first sight appear to demand perfection, they can nevertheless be easily and economically met by a judicious use of the materials at command.

Professor Tyndall has demonstrated that heat will pass with greater ease through a material of uniform density than it will through several mediums of varying density. Arguing from this, the efficiency of insulation should not be considered, as it often is, as depending solely upon the facility with which certain filling materials can resist the passage of heat. It also depends, and to a very large degree, upon the construction of the insulation as regards the external walls, the timber, and the paper. The heat that enters the building comes in chiefly by conduction through the materials forming the wall; the balance comes in by convection and radiation by and through the air. Heat is more easily conducted by a wet body than by a dry one, therefore the insulation should be absolutely dry. In order to convey heat air must move, therefore the insulation should be absolutely air-tight. These two principles are the Alpha and Omega of good insulation, and no matter how it is varied in other particulars they should remain unaltered.

**Air Spaces.**—A study, however superficial, of the American designs for insulation given on pages 223 and 224 will reveal the fact that considerable reliance is placed by American experts on the value of a closed air space for preventing the leakage of heat into a building.

Air spaces are sometimes used in this country between

the inside of the brick wall of the building and the inside layer of matchboarding, the idea being to keep the timber away from the wall so as to lessen the chances of its becoming damped by any moisture coming through the wall.[1] This is a very usual construction, and as the air space in the best designs is not more than $\frac{3}{4}$ of an inch, or, if the grounds are brandered, $1\frac{1}{2}$ to 2 inches, wide, it serves a useful purpose. In fact, it is a case of convenience and cheapness *versus* theoretical efficiency. The insulation must be fixed to something, and, owing to the number required, wood bricks are not satisfactory things to which to fix matchboarding. There is an undoubted advantage in brandering the grounds and so reducing the area of the insulation proper which is in direct contact with the brickwork.

As an insulator *per se*, the air space is, in the opinion of those best qualified to judge, a discredited contrivance. Mr Stoddard[2] has made some interesting experiments for the purpose of demonstrating the value of successive air spaces. These are shown in fig. 68. Mr Stoddard deduces from these experiments that for the purpose of insulation "a wide air space has no greater value than a narrow one, and that any space over one half ($\frac{1}{2}$) inch in width, if it can be kept dry, will be of greater value if filled with an insulating material as good as mill shavings than if left as an air space." The figures on the right of the diagram give the actual space occupied by different designs of insulation

[1] In constructing the cold stores at the Victoria Docks and at West Smithfield, Mr Donaldson purposely left and ventilated a $2\frac{1}{2}$- to 4-inch space between the insulation proper and the wall. The stated object of this was to allow of the free circulation of air and to prevent sweating from the wall. In order to prevent conduction, he placed a $\frac{1}{4}$-inch layer of hair-felt between the wooden grounds and the wall.

[2] *Proceedings American Warehousemen's Association*, 1901–02.

14

having an equal heat-resisting value, but packed with different fillers. Hair-felt, granulated cork, slag wool, and mill shavings were experimented with, and the last example is a series of 6 air spaces.

The actual physical effects of an air space and what takes place in it has been succinctly discussed in *Cold Storage*,[1] from which the following extracts are taken:—

"It is, of course, well known that air spaces in the brickwork are used to keep the damp out of dwelling-

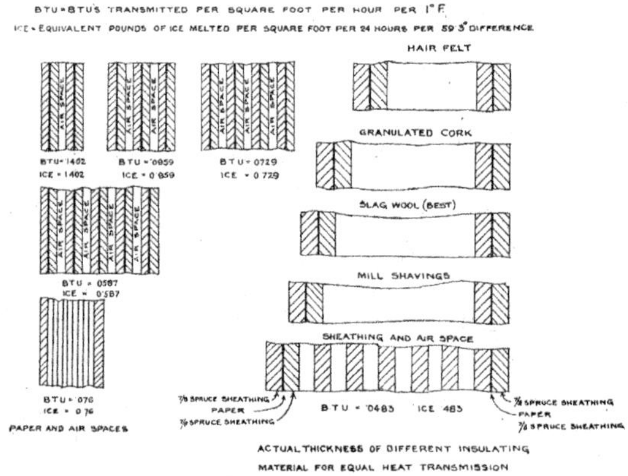


FIG. 68.

houses, especially where the wall of the building is in contact, on its outer side, with the earth, such, for example, as in a cellar, and that in this position air spaces achieve to a greater or less extent their object. It is also well known that with a high temperature difference an air space is useful as a non-conductor, prominent examples being its use as an insulator between the cast-iron of a steam cylinder and the planished iron lagging, or between

[1] *Cold Storage and Ice Trades Review*, July 15th, 1901, p. 93.

the firebrick lining and ordinary brickwork of a furnace chimney shaft. Now, as regards dampness, a cold store is a building of the warehouse class, and, broadly speaking, only differs from an ordinary warehouse in certain details in its construction, and in the fact that a low temperature is required inside it, for which purpose it is insulated. In an ordinary warehouse there is no danger of the percolation of damp through the walls above ground-level, especially if the brickwork is built in cement; and it is generally held that in such buildings air spaces in the brickwork do no good, but only add to the expense and weaken the walls—in short, it is not common or good practice to provide them. If, then, the walls of an ordinary warehouse will keep dry without an air space, and this is a fact beyond dispute, why not those of a cold store? If the wall is below ground-level, the case is different, and there is a likelihood, not only of the percolation of damp, but also of water. In this latter case an air space will lessen, but will not prevent, the evil, and it is necessary either to provide a vertical asphalt damp course with a brick holding-up wall, or to provide an extra 4½ inches of brickwork, which is bonded into the rest of the wall but laid in hot asphalt, thus effectually preventing the ingress of both damp and water.

" As regards the insulating properties of an air space, it is a matter of elementary physics that air, when coming in contact with a warmer surface than itself, will absorb some of the heat, will become light, and will rise; while if it comes in contact with a surface colder than itself, it will give up some of its heat, will become chilled, and fall. Thus, in a confined space, and so long as the temperature difference is maintained, the air will keep up a constant circulation. We may mention, in passing, that the circulation caused by this alternate heating and cooling—with the addition that the air when being cooled deposits its moisture and becomes dried—is the principle of Sir Montague Nelson's defrosting apparatus. It is argued that this is exactly what happens in an air space in insulation, and that, owing to the temperature difference between the outside and inside wall, the outer wall of the air space is warmer than the inner, and, therefore, the air in the air space in contact with the warmer surface becomes warmed and rises until it gets to the top, when, owing to its being confined, it comes in contact with the cooler side, and, giving up its heat, it becomes chilled and falls to the bottom, where it again comes into contact with the warmer wall and repeats the process, thus keeping up a constant circulation and continually conveying heat from the warmer wall to the cooler, and therefore from the outside air to the refrigerating machine. If, on the other hand, the space occupied by the air space were filled with some insulating substance, such as slag wool or charcoal, which contains air in minutely subdivided particles, the insulating properties of the air are obtained without any chance of circulation going on such as we have pictured above."

**Paper.**—It is an essential condition of good insulation that it should be absolutely dry and air-tight. To ensure this, specially made paper of considerable tensile strength, and impregnated with suitable chemicals which render it both air- and moisture-proof, is placed between two or more layers of matchboarding on either side of the filling material. The insulating properties *per se* of paper are not great. It is chiefly valuable in laminating the air, and so producing changes of density in the various portions of the insulation considered as a whole. It is a very cheap substance, and is particularly useful in keeping out moisture and preventing the leakage of air. Reference to the table on page 220 will show that wet shavings will transmit some 55 per cent. more heat than dry shavings, and this is true in a greater or less degree for all fillers. Plain brown paper which has been sometimes used is of no benefit whatever, as it is neither air-tight nor moisture-proof, the two essential properties. Paper for use in insulation should be entirely without smell, waterproof, air-tight, strong enough to bear handling without tearing, and, of course, of a lasting quality. When erected it should be either laid in one double layer, breaking joint into each other, between two thicknesses of matchboarding, or in two single layers between three thicknesses of matchboarding. Particular care must be taken with the corners of the rooms and the junctions of the floors and walls, and the paper should be carefully fixed and the joints lapped. Too much trouble cannot be taken with this material, as upon its being perfect depends to a very large degree the success of the insulation. It should always be close laid between timbers. It is not advisable to lay it only on laths as is sometimes done, or stretch it between studs or joists with timber support on one side only, as shown in fig. 68.

**Tests of Insulation and Filling Material.**—Numerous tests have from time to time been made by various authorities to endeavour to ascertain what material or combination of materials possessed the property of resisting the passage of heat in the most marked degree.

In 1899 Messrs Lamb & Wilson carried out some tests of filling material, the results of which were published in the *Proceedings of the Royal Society.* The substances to be tested were placed in a space between two cylindrical copper pots, which were kept apart and at a fixed distance from each other by pieces of vulcanized fibre. The inner pot contained an electric heating coil and a small motor with a fan attached, the object of this latter being to guide the currents of air evenly over the internal surface of the copper pot. The leads supplying electric energy to the motor and coil which were run in series were carried through holes in the lid of the pots, and as the current and voltage was measured, the energy consumed could be accurately estimated.

Any heat which filtered through the filling material was carried away by a stream of cold water in which the outer pot was immersed, and the temperature difference was measured by a thermo-electric indicator, and varied from 8° to 28°. The following specific results were obtained[1]:—

| Material. | Conductivity. |
|---|---|
| Air (no baffles), | 0·000200 |
| Pine sawdust, | 0·000242 |
| Pine shavings, | 0·000162 |
| Brown paper crumpled up, | 0·000167 |
| Hair-felt broken up and carded, | 0·000145 |
| Hair-felt in two sheets each ½ in. thick, | 0·000106 |
| Dry asbestos, | 0·000297 |
| Charcoal, | 0·000150 |
| Sand, | 0·000740 |
| Rice husks, | 0·000150 |
| Kapok, a heat insulator, | 0·000144 |
| Kapok, loose, | 0·000122 |
| Silicate cotton, | 0·000151 |

These results give the best value for hair-felt. Char-

[1] *Proceedings Royal Society*, 1899, pp. 283, 288.

coal, rice husks, and silicate cotton are (within the error of the apparatus) an equal second. Cork was not tested.

In 1897 Mr F. H. Donaldson read a paper before the Institution of Civil Engineers,[1] in which he described some tests he had carried out on various filling materials. For the purpose he had had a number of tin boxes made. These were provided with drain pipes, and were as nearly as possible similar. In these boxes were placed blocks of ice of the same weight, and the box was embedded in the material to be tested. At the end of a certain time the boxes were opened and the ice weighed. The filling material which retained the largest block of ice was held to be the best. The following table of results, which have been extended as regards the last column by Mr Bost, are taken from the *Proceedings of the Cold Storage and Ice Association*:—

EXPERIMENT No. 1.

| Material tested. | Thickness of insulating material. | Original weight of ice. | Weight after | | Loss after 72 hrs. | Oz. per external sq. foot per hour. |
|---|---|---|---|---|---|---|
| | | | 24 hours. | 72 hours. | | |
| | in. | oz. | oz. | oz. | p.c. | oz. |
| Peat (compressed and set in fossil meal), | 9 | 95 | 81 | 59 | 37·89 | ... |
| Charcoal, . . | 11 | 96½ | 79½ | 56 | 41·97 | ·0062 |
| Silicate cotton, . | 4½ | 92½ | 73½ | 40½ | 56·21 | ·0072 |
| Magnesia and asbestos fibre, . . | 4½ | 93 | 73 | 40½ | 56·45 | ... |

[1] *Proceedings Inst. C.E.*, vol. cxxix. p. 1.

## Experiment No. 2.

| Material tested. | Thickness of insulating material. | Original weight of ice. | Weight after | | | Loss after 96 hrs. | Oz. per external sq. foot per hour. |
|---|---|---|---|---|---|---|---|
| | | | 24 hours. | 48 hours. | 96 hours. | | |
| | in. | oz. | oz. | oz. | oz. | p.c. | oz. |
| Silicate cotton, . | 6 | 104 | 88¾ | 76¾ | 58½ | 43·75 | ·0071 |
| Sawdust, . . | 9 | 103½ | 86½ | 71 | 48 | 53·62 | ... |
| Peat, . . . | 9 | 104 | 77½ | 56 | 26 | 74·75 | ... |
| Charcoal, . . | 9 | 104 | 88¾ | 78½ | 60½ | 41·82 | ·0067 |

## Experiment No. 3.

| Material tested. | Thickness of insulating material. | Original weight of ice. | Weight after | | Loss after 72 hrs. | Oz. per external sq. foot per hour. |
|---|---|---|---|---|---|---|
| | | | 24 hours. | 72 hours. | | |
| | in. | oz. | oz. | oz. | p.c. | oz. |
| Silicate cotton, . | 9 | 92 | 83½ | 72½ | 21·19 | ·019 |
| Charcoal, . . | 11 | 92 | 82¾ | 70½ | 23·36 | ·012 |

## Experiment No. 4.

| Material tested. | Thickness of insulating material. | Original weight of ice. | Weight after 24 hours. | Weight after 96 hours. | Loss after 96 hrs. |
|---|---|---|---|---|---|
| | in. | oz. | oz. | oz. | p.c. |
| Silicate cotton (loosely packed), . . . | 9 | 110 | 103 | 84¼ | 23·41 |
| Silicate cotton, . . . | 9 | 110 | 101¾ | 80¾ | 26·59 |
| Charcoal, . . . . | 11 | 110 | 100½ | 79 | 28·18 |
| Vegetable silica, . . | 11 | 110 | 101½ | 76¾ | 30·22 |
| Diatomite, . . . . | 11 | 110 | 99 | 73¾ | 32·95 |

In some further tests that were recently made, two boxes, one smaller than the other, were put one inside the other. The size of the inside box was such as to leave a space of some four inches between the two. This space was filled with the material to be tested, and a carefully calibrated maximum and minimum thermometer was put inside the inner box. Both boxes were placed in a temperature of 23° F., and after being thoroughly chilled were sealed up for 36 hours. They were then taken out and placed in a temperature of 50° F. After 24 hours in this temperature the boxes were quickly opened and the rise in temperature noted. A number of tests were made, the different boxes being packed with different materials.

The results were as follows:—

| Material tested. | Rise in degs. F. after 24 hours. | Weight per cubic foot. |
|---|---|---|
| Silicate cotton, . . | 14 | 13·5 |
| Flake charcoal, . . | 14·5 | 15 |
| Granulated cork, . . | 15 | 8 |
| Sawdust (deal), . . | 15·5 | 15·75 |
| Twig charcoal, . . | 16·5 | 10·5 |
| Shavings, . . . | 16·5 | 4·5 |
| Mahogany (dust), . . | 16·5 | 16·5 |

Other tests have from time to time been made, but the results are mostly paraphrases of the same story. No one seems inclined to take anyone else's tests as final or conclusive, and this is probably wise. It would, however, be of great value if, as previously suggested, tests of different insulations as a whole, and not the filling material alone, could be made, and the results published showing the heat leakage for different designs per square foot with different temperature heads. Mr Stoddard, whose valuable researches have previously been referred to, has done this to a certain degree. His results are set out in fig. 69 and the table following.[1] Fig. 69 shows a section of a wall. The brickwork is 24 inches thick, and against this on the inside is a 2-inch air space, then two layers of ⅞-inch tongued and grooved spruce with paper between, then 12 inches of spruce mill shavings and similar timber and paper on the inside. The external temperature was assumed to be 92° F., which is high, and the temperature in the store 32° F. The

[1] *Proceedings American Warehousemen's Association*, 1901–02.

method of heat transmission through this, as, indeed, through any insulator, is as follows:—The heat is conveyed to the brickwork by radiation from the sun and air, and passes through the brickwork by conduc-

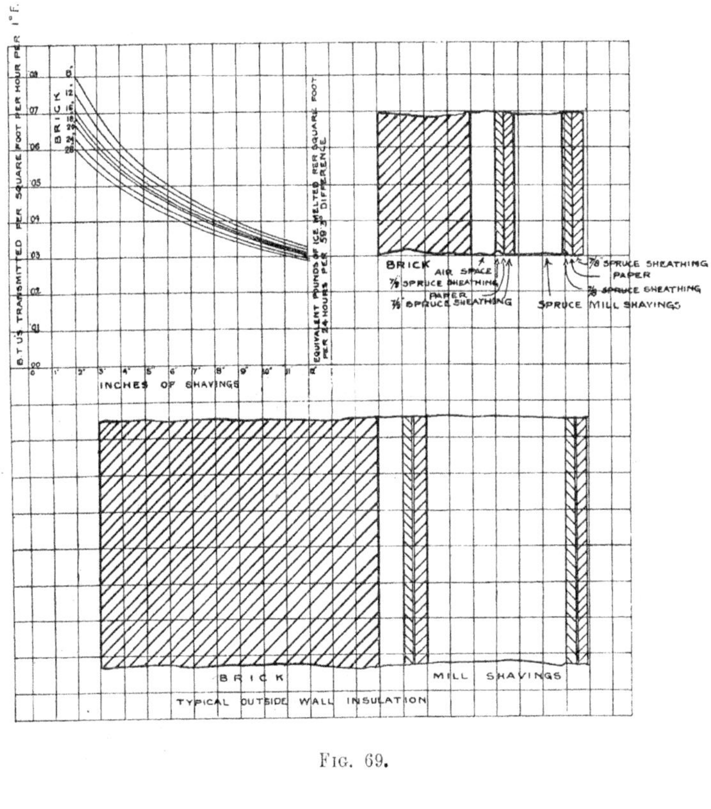


FIG. 69.

tion. It then passes across the air space, and is finally disseminated from the inside surface by radiation and convection.

In diagram fig. 70 the B.T.U. transmitted per square foot per hour through an ordinary stud partition consisting of two layers of $\frac{7}{8}$-inch tongued and grooved spruce sheathing, with paper between, on each side of the studs, the filling material of spruce mill shavings varying in thickness from 2 to 24 inches, are given.

TABLE showing transmission of heat through various insulating structures.

Col. 1 gives B.T.U. per square foot per day per degree of differences of temperature. Col. 2 gives meltage of ice in lbs. per day by heat coming through 100 square feet at a difference of 40°.

| | Col. 1. | Col. 2. |
|---|---|---|
| $\frac{5}{8}$" oak-paper, 1" lamp-black, $\frac{7}{8}$" pine (this is the ordinary small stock family refrigerator), . . | 5·7 | 160·7 |
| One $\frac{7}{8}$" board, 1" pitch, one $\frac{7}{8}$" board, . . . . | 4·90 | 138 |
| Four $\frac{7}{8}$" spruce boards, two papers, solid, no air space, | 4·28 | 120 |
| Two double boards and paper (four $\frac{7}{8}$" boards), and one air space, . . . . . . . . . | 3·71 | 105 |
| One $\frac{7}{8}$" board, 2" pitch, one $\frac{7}{8}$" board, . . . . | 4·25 | 119·7 |
| One $\frac{7}{8}$" board, $2\frac{1}{2}$" mineral wool, paper, one $\frac{7}{8}$" board, | 3·62 | 101·9 |
| Two $\frac{7}{8}$" double boards and two papers, 1" hair-felt, . | 3·318 | 93·4 |
| Two $\frac{7}{8}$" boards and paper, 1" sheet-cork, two $\frac{7}{8}$" boards and paper, . . . . . . . . | 3·30 | 92·9 |
| One $\frac{7}{8}$" board, paper, 2" calcined pumice, paper and $\frac{7}{8}$" board, . . . . . . . . . . | 3·38 | 95·2 |
| Four double $\frac{7}{8}$" boards with paper between (eight boards) and three 8" air spaces, . . . . . | 2·7 | 76 |
| Hair quilt insulator, four boards, four quilts hair, . | 2·517 | 70·9 |
| One 7" board, 6" pat. silicated strawboard, air cell finished inside with thin layer pat. cement, . | 2·48 | 69·8 |
| One $\frac{7}{8}$" board, paper, 3" sheet-cork, paper, one $\frac{7}{8}$" board, . . . . . . . . . . | 2·10 | 60 |
| Two $\frac{7}{8}$" boards and paper, 8" mill shavings and paper, two $\frac{7}{8}$" boards and paper, . . . . . . | 1·35 | 38·3 |
| Same, slightly moist, . . . . . . . | 1·80 | 50·7 |
| Same, damp, . . . . . . . . . . | 2·10 | 60 |
| Double boards and paper, 1" air, 4" sheet-cork, paper, one $\frac{7}{8}$" board, . . . . . . . | 1·20 | 33·6 |
| Same, with 5" sheet-cork, . . . . . . | ·90 | 25·3 |
| $\frac{7}{8}$" board, paper, 1" mineral wool, paper, $\frac{7}{8}$" board, . | 4·6 | 130 |
| Double boards and papers, 4" granulated cork, double boards and paper, . . . . . . | 1·7 | 48 |

## DIFFERENT METHODS OF CONSTRUCTION IN VOGUE.

It would be a difficult task to find a subject of design with which fancy had played more pranks or had wandered with a more unchecked hand than in the designs for insulation. With the exception of double walls, or walls built with a hollow brick or tile lining—the space thus

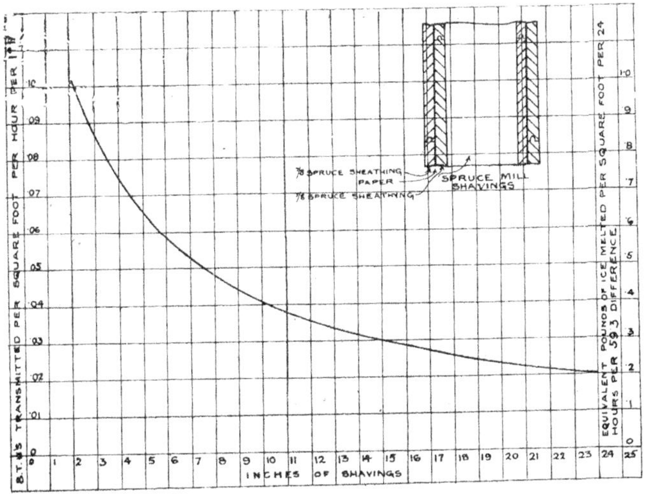


FIG. 70.

formed being sometimes filled with bitumen or pitch and sometimes left plain—the two methods of construction most in vogue, and round which the most discussion wages, are those with and without air spaces. Some authorities discard air spaces altogether, and make up the insulation by alternate layers of paper and matchboard; others, while adopting air spaces, differ greatly as to the manner in which they should be applied; again, some advocate two or three courses of timber and paper, and one or two air spaces sometimes separated from each other by timber and matchboarding, and sometimes by paper secured to battens alone. Then comes the design of matchboard and paper partitions with one or more air spaces filled with some insulating material, and lastly there is the design of paper and matchboard partition with the air spaces left empty but divided up into many squares, like a chess-board, by cross timbers between the studs. Designs for insulation can be extravagant (1) in space, (2) in timber, and (3) in labour. The best is a mean between these three. In all cases, and no matter which of the above are used, the two things previously mentioned stand essential:—(1) that whether the air spaces are empty or filled, they must be enclosed in partitions which are as air-tight as skill and good material can make them; and (2) that they—and if they contain filling material, it—must be perfectly dry.

**Outer Walls.**—The outside wall of the building should be coated after completion with air and waterproof[1] enamel paint, white for preference. If the use of paint is objected to, whitewash[2] can be used. It will prove much cheaper, and, if properly made, equally effective for standing the weather. The inside of the wall should be coated with pitch or tar of such a mixture that it will not crack when the inside temperature is reduced to the required degree. Bitumastic solution applied as paint is useful for this purpose. By confining the air

[1] See paper by Mr Stoddard read before the American Warehousemen's Convention (*Proceedings*, 1901).

[2] Weather-proof whitewash can be made by mixing ordinary whitewash in water and adding salt in the proportion of about 50 lbs. of salt to 40 gallons of whitewash. Another receipt says mix six parts of lime in warm water to the required strength, then add three parts of salt when solution has taken place, and just before use add one pint of Portland cement.

contained in the brick wall between two impenetrable coatings in this way, it is kept from circulating and becomes dead, and in the best state to act as an insulator. Sometimes the outer walls are built with the outside 4½ inches made of hollow tiles. These are laid so that the holes in them form a continuous opening from the bottom of the wall to the top. This allows the admission of cold air at the bottom and its escape warm at the top, the object being to thus carry off any heat which may by conduction or radiation have passed through the white-wash and stucco. Most authorities are fairly well agreed on the treatment of the outer wall, the exception being those who leave it severely alone. The real variations begin when the inner insulation is being considered. A few illustrations, taking American practice first, may make this clearer.

In one design, against the inside of the wall, but secured to it, are fixed 2-inch by 4-inch studs. On these are nailed double matchboarding, the two layers breaking joint with waterproof paper between them. The 4-inch space between the wall and the first layer of matchboarding is broken up or shelved with timber placed between the joists about 12 inches one above the other, the object being to provide a number of air spaces of small area. Sometimes these spaces are filled with insulating material, sometimes not. If the outside of the wall cannot for any reason be protected in the manner already described, there may be a series of two or more of these air spaces. In another design 2-inch by 2-inch studs are fixed against the inside wall, and then a number, say three 2-inch air spaces, are formed, the partitions between these being composed of matchboarding in two layers laid to break joint. Between the two courses of matchboarding is laid a two-ply waterproof impregnated paper. The air spaces are sometimes left plain, and sometimes some or all of them are filled with some good filling material. Sometimes the two outside spaces are left empty and the inside space filled, sometimes the reverse. In both these cases the layer of matchboarding on the inside of the room is thicker than the other, and is sized and twice varnished. Mr Skinkle[1] has given a number of illustrations of different designs of insulations which may be taken as typical of American practice. Some of these are reproduced in fig. 71.

British practice in this matter has also varied largely, but, contrary to the American, the general tendency has been to use a great amount of filling material and almost entirely to ignore air spaces, except a small one, placed immediately against the brick wall, to guard against dampness. Fig. 72 illustrates a very good design. The inside of the wall above ground-level is painted with a pitch solution to keep out any moisture that might find its way through. Below ground-level it is protected from moisture by a $\frac{3}{4}$-inch layer of asphalt put on in two layers so as to prevent the possibility of pin holes being left, the whole being held up to its work and kept in place by a 9-inch lining wall laid in cement. On the inside of the wall and of the lining wall 2-inch by $\frac{3}{4}$-inch rough battens are fixed. On these, two layers of $\frac{1}{2}$-inch matchboarding with double waterproof paper between them are laid, each layer breaking joint with the one before it. Then 6-inch by 2-inch studs are fixed, and on the inside of these another layer of $\frac{1}{2}$-inch matchboarding is laid. Then another layer of paper, another layer of $\frac{1}{2}$-inch matchboarding, another layer of paper, and finally inside all a

[1] *Ice and Refrigeration*, Chicago, Nov. 1899.

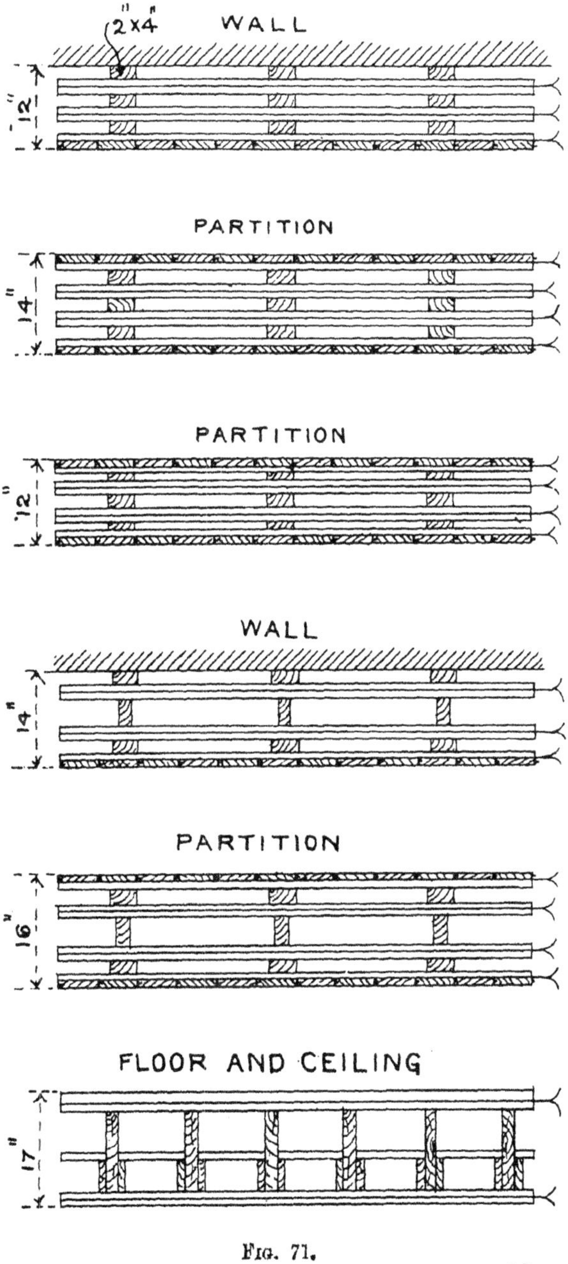

2"x4"
WALL
12"
PARTITION
14"
PARTITION
12"
WALL
14"
PARTITION
16"
FLOOR AND CEILING
17"


Fig. 71.

layer of 1-inch matchboarding protected from damage by 2-inch by 2-inch battens. The inside wall can be whitewashed, or, if it is preferred, varnished. The 6-inch space is filled with charcoal or silicate cotton. The design gave good results in practice. No special precaution was taken to keep the external face of the wall

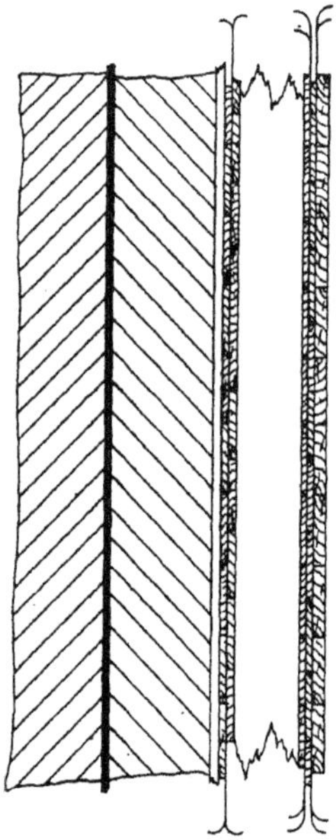

FIG. 72.

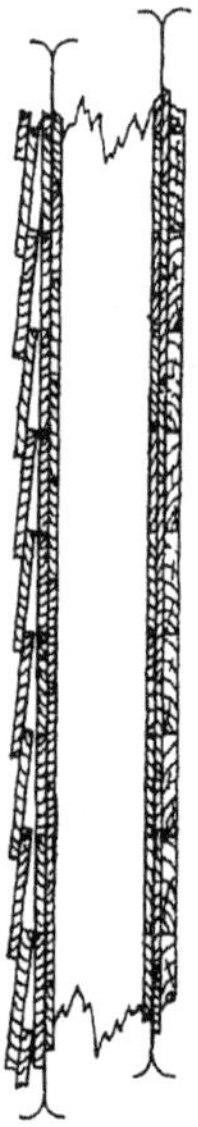

FIG. 73.

FIG. 74.

above ground-level air- or water-tight, but no trouble from dampness was experienced.

In freezing works in the Colonies, where the buildings are built entirely of wood, the construction shown in fig. 73 is usual. The main wall is formed by 2-inch by 6-inch studs placed at 12-inch centres or nearer. On the weather side of these, two layers of ½-inch matchboarding breaking joint are laid with paper between, and on these again the ordinary weather-boarding is nailed. Sometimes the external ½-inch matchboarding gives place to 1-inch flush weather-boarding, with special weather-resisting rebate.

The inside of the 6-inch by 2-inch studs is lined with a layer of ½-inch matchboarding, then a layer of paper, then another layer of ½-inch matchboarding, another layer of paper, and, finally, inch boarding, the inside of which is limewhited. In some cases a 2-inch air space has been placed between the first and second layer of matchboarding, but in more recent practice this has been omitted. In all cases the spaces between the 6-inch by 2-inch studs is filled with sawdust, calcined pumice, cowhair, or some other local and cheap filling material. In the early days, before the introduction of paper, the same form of construction was used omitting this, but the results obtained were, of course, not so good.

**Partitions.**—The best construction for the partitions depends, as in the case of floors, on the temperature difference; if this is small, the flow of heat will be comparatively trifling, and for this reason it is not necessary to expend large sums of money in insulating rooms from each other, the atmosphere of which does not differ many degrees in temperature. All that is generally necessary in these cases is to have an ordinary stud partition, the space between the studs being either left blank or filled with some comparatively cheap insulator, the studs being covered on each side with double matchboarding with paper between. In some cases double matchboarding and paper are put on one side only, the other side being covered with single matchboarding; and in extreme cases the paper is omitted altogether, single matchboarding on each side only being fixed. Fig. 74 illustrates a good partition.

**Floors.** — Floors have not received quite so much attention as walls, because up to the present in this country most of the larger stores have been designed for uniform temperature, and it has been the exception to insulate the intermediate floors. However, with the demand for cold stores having rooms capable of being maintained at any required temperature, the case has

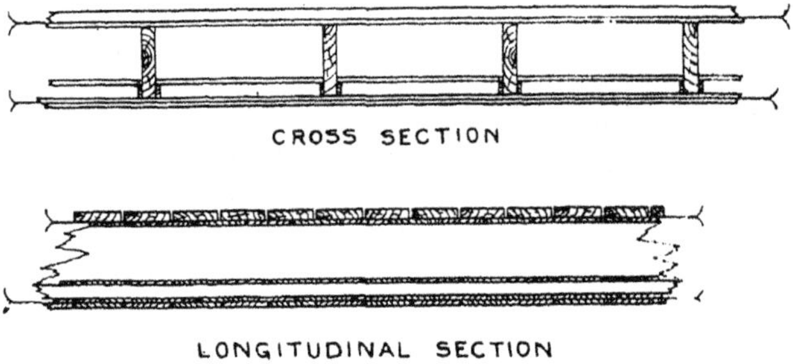


FLOOR

FIG. 75.

altered, and floors and partitions or bulkheads have to be insulated. Naturally the extent and type of the insulation adapted depend on the temperature defence between the room above and that below, and also on what class of goods are likely to be stored.

Fig. 75 illustrates a very good design, which, though rather expensive in first cost, is quite waterproof and a good insulator.

The construction is as follows:—

The 9-inch by 2-inch joists are covered on the underside

with two thicknesses of ½-inch matchboarding with one or two layers of waterproof impregnated paper between them. The space between the joists is three-quarters pugged with some good filling material (it can be quite filled if desired), and the joists are floored with a layer of ½-inch or ¾-inch matchboarding. On this is laid one or two layers of ruberoid, or, rather, heavy waterproof or Willesden roofing paper, the top layer breaking joint with the one below it. This paper covers the whole area of the floor, and is flashed up to the paper in the walls on each side to a height of at least 12 inches. It should be carefully secured at all the joints, and care must be taken that it is absolutely waterproof. On top of this paper is laid plain flooring timber—*i.e.*, not tongued and grooved—the boards being not more than 6 inches wide or less than 1½ inches thick. These boards must not be cramped up, and, when laid, they must have their joints calked with oakum and run in with pitch in a manner similar to a ship's deck. A heavy skirting-board, 6-inch by 2-inch or 9-inch by 2-inch, is housed to the floor and secured to the insulation of the wall of the building. The arris must be chamfered off, and the joint between the underside of the skirting-board and the floor must be calked and pitched like the floor. This construction is, as before stated, absolutely air- and water-tight, and can be scrubbed down and cleaned off whenever necessary. There is no fear of damaging either the insulators or the goods in the room below by storing ice or soft goods liable to give off moisture, and the timber floor will stand considerable wear and tear.

A lighter floor, which is used where dampness has not to be guarded against to the same extent, consists merely of two thicknesses of tongued and grooved matchboarding with waterproof paper between as a ceiling. And as a floor, first a thickness of ½-inch or ¾-inch timber, then a layer of paper, and then 1¼ tongued and grooved flooring, iron tongues being used. The space between the joists is three-quarters pugged, forming a 6-inch thickness of filling material.

## CHAPTER IX.

### BRINE.

**Sodium Chloride** is a mixture of common salt and water. The object of using brine in connection with refrigerating plants is, of course, to obtain a liquid capable of carrying a lower temperature than 32° F. without freezing. Both chloride of sodium and chloride of calcium are used for this purpose, and they both answer equally well within ordinary temperature ranges. Owing, however, to certain properties touched on later, the latter is generally preferred. The table[1] on the following page gives the properties of sodium chloride solution. In the second and fourth columns the U.S. gallon of 8⅓ lbs. is referred to, and it has not been thought worth while to alter them to the Imperial standard.

An examination of the table shows that as the strength of the solution increases, the specific heat falls, so that it is a direct gain not to use a stronger solution than will withstand, without freezing, the lowest temperature it is likely to meet. A property possessed by sodium chloride, which is not shared by its rival, is that when too strong a solution is used, the salt is apt to deposit in the brine-tanks and coils, choking them and stopping the circulation.

[1] Siebel, *Compend of Mechanical Refrigeration.*

But a greater danger than this, and one shared equally by both, is that if the solution is not strong enough, or for some reason or other the back pressure is carried too low, the brine will freeze on the expansion coils of the cooler and prevent them working. The brine supply freezing the water in an ice skating-rink was once stopped in this way. The calcium chloride froze solid on the cooling coils, which were of the Baudelot type; the ice surface refused to form, and the public left in disgust. When the fault was discovered, it was found to be due to a new engine-driver, who had closed his expansion cocks too much, and was carrying a very low back pressure, and so temperature, in the coils.

TABLE SHOWING PROPERTIES OF SOLUTION OF SALT. (CHLORIDE OF SODIUM.)

| Percentage of salt by weight. | Pounds of salt per gallon of solution. | Degrees on Salometer at 60° F. | Weight per gallon at 39° F. = 4° C. | Specific gravity at 39° F. = 4° C. | Specific heat. | Freezing point, Fahrenheit. | Freezing-point, Celsius. |
|---|---|---|---|---|---|---|---|
| 1 | 0·084 | 4 | 8·40 | 1·007 | 0·992 | 30·5 | - 0·8 |
| 2 | 0·169 | 8 | 8·46 | 1·015 | 0·984 | 29·3 | - 1·5 |
| 2·5 | 0·212 | 10 | 8·50 | 1·019 | 0·980 | 28·6 | - 1·9 |
| 3 | 0·256 | 12 | 8·53 | 1·023 | 0·976 | 27·8 | - 2·3 |
| 3·5 | 0·300 | 14 | 8·56 | 1·026 | 0·972 | 27·1 | - 2·7 |
| 4 | 0·344 | 16 | 8·59 | 1·030 | 0·968 | 26·6 | - 3·0 |
| 5 | 0·433 | 20 | 8·65 | 1·037 | 0·960 | 25·2 | - 3·8 |
| 6 | 0·523 | 24 | 8·72 | 1·045 | 0·946 | 23·9 | - 4·5 |
| 7 | 0·617 | 28 | 8·78 | 1·053 | 0·932 | 22·5 | - 5·3 |
| 8 | 0·708 | 32 | 8·85 | 1·061 | 0·919 | 21·2 | - 6·0 |
| 9 | 0·802 | 36 | 8·91 | 1·068 | 0·905 | 19·9 | - 6·7 |
| 10 | 0·897 | 40 | 8·97 | 1·076 | 0·892 | 18·7 | - 7·4 |
| 12 | 1·092 | 48 | 9·10 | 1·091 | 0·874 | 16·0 | - 8·9 |
| 15 | 1·389 | 60 | 9·26 | 1·115 | 0·855 | 12·2 | -11·0 |
| 20 | 1·928 | 80 | 9·64 | 1·155 | 0·829 | 6·1 | -14·4 |
| 24 | 2·376 | 96 | 9·90 | 1·187 | 0·795 | 1·2 | -17·1 |
| 25 | 2·488 | 100 | 9·97 | 1·196 | 0·783 | 0·5 | -17·8 |
| 26 | 2·610 | 104 | 10·04 | 1·204 | 0·771 | -1·1 | -18·4 |

Sodium chloride can be made in any simple way. The same remarks apply to calcium chloride. Avoid putting the salt into the ice-tanks and trying to mix it there. It will be exceedingly difficult, and the salt will cake on the bottom of the tank and on the lower expansion pipes. The brine should always be mixed in a separate vessel, and then put into the tanks. An ordinary way of doing this is to make a false bottom to a common barrel or box, perforate this with small auger holes, and fix it some three or four inches above the real bottom. Put the salt or chloride into the barrel and fill it nearly to the top. Connect up the bottom portion with a hose, turn on the water, and let it rise up through the salt. To rid it of any impurities or undissolved salt, the brine should be strained as it runs out at the top through a pipe in the side of the barrel provided for that purpose. The water supply to the barrel should be capable of regulation, so as to prevent the barrel overflowing. It is best to mix brine rather stronger than is wanted, and dilute it down to the proper strength by the subsequent addition of water.

**Calcium Chloride** is a bye-product of chemical works, and is obtained in large quantities in the manufacture of salt, $CO_2$, and ammonia. Generally speaking, chloride of calcium and chloride of sodium are similar in character, but the former is the more energetic moisture-absorber, for, whereas salt will only absorb sufficient moisture to make it damp, chloride of calcium will absorb nearly its own weight of water and become liquid. It has accordingly been used to a great extent as a moisture-absorber in connection with processes requiring a dry atmosphere. One great objection to sodium chloride brine is that it rusts the ice-tanks and coils immersed in it and causes considerable wear. This is avoided by using calcium chloride. Sodium brine at its maximum density will freeze at about $-7°$ F., while calcium brine can be made, which, if necessary, will not freeze until it reaches a temperature of $-50°$ F. The specific heat of calcium brine is higher than that of sodium brine, so that it is not necessary to circulate so much of it through the system to produce a certain refrigerating effect, and, as before stated, it is not so apt to deposit and crystallize in the pipes.

A 20 per cent. solution of calcium chloride is quite dense enough, and has a sufficiently low freezing-point for all practical purposes. If specially low temperatures are required, a 25 per cent. solution may be used. It is desirable, however, not to have too strong a solution, and so long as the brine is sufficiently dense to prevent the liquid freezing on the expansion coils, nothing further is required. Commercial calcium chloride as sold in drums contains about 25 per cent. of water, and this has to be allowed for in calculating the mixtures.

Salometers as usually sold are correct at a temperature of 60° F., and the brine must be either warmed up or cooled down to this temperature before a proper test of its density can be made.

The tables[1] on the two following pages by Mr Madison Cooper are approximately accurate and give the properties of calcium chloride brine at various densities; the smaller table has, for greater utility, been altered approximately from U.S. gallons of $8\frac{1}{3}$ lbs. to Imperial gallons of 10 lbs.

In preparing brine from calcium chloride, the method usually adopted is to mix the calcium and water in a tank in such proportions as will give a higher density than that actually required. The mixture should be stirred round

[1] *Ice and Refrigeration*, Chicago, May 1900.

## Properties of Solution of Chloride of Calcium. (Calcium Brine.)

| Degrees on Salometer, 60° F. | DegreesBeaumé, 60° F. | Specific gravity, 60° F. | Per cent. of chloride of calcium. | Freezing-point, degrees F. | Freezing-point, degrees Celsius. | Ammonia gauge. | Specific heat. |
|---|---|---|---|---|---|---|---|
| | | | | | | Pressures. | |
| 4 | 1 | 1·007 | 1 | +31·10 | − ·5 | 46 | ·996 |
| 8 | 2 | 1·015 | 2 | +30·38 | − ·9 | 45 | ·988 |
| 12 | 3 | 1·024 | 3 | +29·48 | − 1·4 | 44 | ·980 |
| 16 | 4 | 1·032 | 4 | +28·58 | − 1·9 | 43 | ·972 |
| 22 | 5·5 | 1·041 | 5 | +27·68 | − 2·4 | 41·5 | ·964 |
| 26 | 6·5 | 1·049 | 6 | +26·60 | − 3·0 | 39·5 | ·960 |
| 32 | 8 | 1·058 | 7 | +25·52 | − 3·6 | 38 | ·936 |
| 36 | 9 | 1·067 | 8 | +24·26 | − 4·3 | 37 | ·925 |
| 40 | 10 | 1·076 | 9 | +22·8 | − 5·1 | 35·5 | ·911 |
| 44 | 11 | 1·085 | 10 | +21·3 | − 5·9 | 34 | ·896 |
| 48 | 12 | 1·094 | 11 | +19·7 | − 6·8 | 32·5 | ·890 |
| 52 | 13 | 1·103 | 12 | +18·1 | − 7·7 | 30·5 | ·884 |
| 58 | 14·5 | 1·112 | 13 | +16·3 | − 8·7 | 28 | ·876 |
| 62 | 15·5 | 1·121 | 14 | +14·3 | − 9·8 | 26 | ·868 |
| 68 | 17 | 1·131 | 15 | +12·2 | −11·0 | 23·5 | ·860 |
| 72 | 18 | 1·140 | 16 | +10·0 | −12·2 | 21·5 | ·854 |
| 76 | 19 | 1·150 | 17 | + 7·5 | −13·6 | 20 | ·849 |
| 80 | 20 | 1·159 | 18 | + 4·6 | −15·2 | 18 | ·844 |
| 84 | 21 | 1·169 | 19 | + 1·7 | −16·8 | 15 | ·839 |
| 88 | 22 | 1·179 | 20 | − 1·4 | −18·6 | 12·5 | ·834 |
| 92 | 23 | 1·189 | 21 | − 4·9 | −20·5 | 10·5 | ·825 |
| 96 | 24 | 1·199 | 22 | − 8·6 | −22·6 | 8 | ·817 |
| 100 | 25 | 1·209 | 23 | −11·6 | −24·8 | 6 | ·808 |
| 104 | 26 | 1·219 | 24 | −17·1 | −27·3 | 4 | ·799 |
| 108 | 27 | 1·229 | 25 | −21·8 | −29·9 | 1·5 | ·790 |
| | | | | | | Vacuum | |
| 112 | 28 | 1·240 | 26 | −27·0 | −32·8 | 1 | ·778 |
| 116 | 29 | 1·250 | 27 | −32·6 | −35·9 | 5 | ·769 |
| 120 | 30 | 1·261 | 28 | −39·2 | −39·6 | 8·5 | ·757 |
| ... | 31 | 1·272 | 29 | −46·3 | −43·5 | 12 | ... |
| ... | 32 | 1·283 | 30 | −54·4 | −48·0 | 15 | ... |
| ... | 33 | 1·294 | 31 | −52·5 | −46·9 | 10 | ... |
| ... | 34 | 1·305 | 32 | −39·2 | −39·6 | 4 | ... |
| ... | 35 | 1·316 | 33 | −25·2 | −31·8 | 1·5 | ... |
| ... | 35·5 | 1·327 | 34 | − 9·7 | −23·2 | ... | ... |
| ... | 36·5 | 1·338 | 35 | + 2·8 | −16·2 | ... | ... |
| ... | 37·5 | 1·349 | 36 | +14·3 | − 9·8 | ... | ... |

PRACTICAL TABLE FOR MAKING CALCIUM BRINE.

| Pounds chloride of calcium (commercial fused) to one gallon of water. | Degrees Salometer, 60° F. | Degrees Beaumé, 60° F. | Freezing-point, degrees F. |
|---|---|---|---|
| 3 | 80 | 20 | 4 |
| 3¾ | 88 | 22 | − 2 |
| 4½ | 96 | 24 | − 9 |
| 5 | 104 | 26 | −17 |
| 5½ | 112 | 28 | −27 |
| 6 | 120 | 30 | −39 |
| 7 | ... | 32 | −54 |

until solution is complete. It is best to avoid putting the calcium in the ice-tank, though this is not so bad as in the case of salt. Mixing will be facilitated if the water is warmed up by a steam-coil.

**Regenerating Brine.**—As the brine, more particularly in air-blast plants, is constantly absorbing moisture from the atmosphere, it is necessary, in order to keep it at its proper strength, to either regenerate it by adding more calcium or salt, or to concentrate it by evaporating off some of the absorbed moisture. The latter is the more useful method, for, if the brine was always being regenerated by the addition of fresh calcium or sodium, the supply of brine in the system would be constantly increasing until more was obtained than was really necessary.

**Concentrating Brine.**—A simple apparatus for concentrating brine and bringing its strength up to whatever density is required, consists of a steam-jacketed pan, very similar to those used for jam-boiling or lard-melting.

The brine, as it requires to be concentrated, is run into the pan from the battery tank. Live steam is admitted into the jacket and the brine in the copper boils, the excess water being driven off as steam. After the brine has been concentrated sufficiently and has cooled down, it

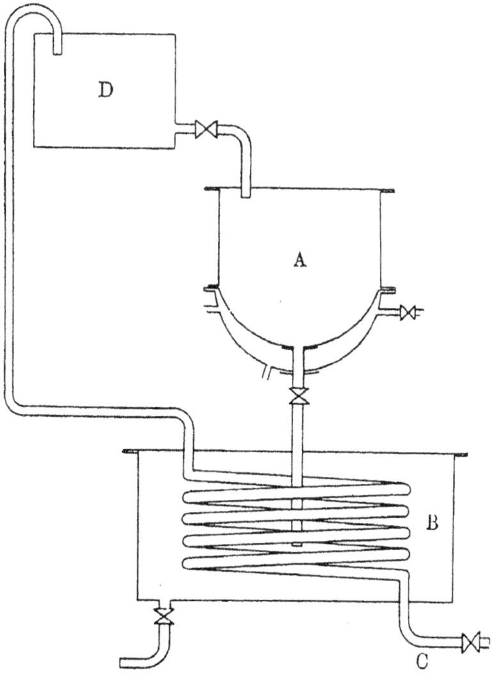


FIG. 76.

is drained off to the brine pump and reabsorbed into the system.

In large plants where a considerable amount of moisture is absorbed by the brine as the air passes over the battery, the brine has to be frequently concentrated. In this case the method shown in fig. 76 [1] is used. In the diagram, *A* is a steam-jacketed copper pan similar to that just described; *B* is a tank which contains the hot brine just concentrated, and *D* is a warm brine tank. When the apparatus is at work, a charge of cold brine is put into *A* and concentrated to the proper density; it is then discharged hot into *B*. Cold brine requiring to be concentrated flows from the battery tank through the pipe *C* and through the coils in *B* into the tank *D*. In passing through the coils it is warmed by the hot brine in the tank, and when it enters *D* is already hot. The concentrated brine is correspondingly cooled. The coil in the tank *D* thus becomes a heat interchanger and saves fuel by warming up the cold brine. This interchanger may be of the open-air type, and if this is adopted, the hot brine is further cooled by the action of the atmosphere.

The cooler is usually just below the level of the battery tank, and the brine is gravitated through the coil in *B* into the tank *D*, the supply being regulated by the valve shown. In this case it has, of course, to be returned to the tank by a pump. When the concentrator is not in continuous use the tank *D* may be omitted. In practice the copper *A* is fitted with a bonnet to catch and carry off the steam.

[1] Messrs the Haslam Foundry and Engineering Company, Ltd., Derby.

# CHAPTER X.

## DISTILLED WATER.

**Introductory.**—It is only in recent years that ice, in this country, has been used to any extent for table purposes, and, the demand being small, the imported clear natural ice has been sufficient to supply all wants. Ice drawn from the deep Norwegian or Canadian lakes is probably as pure as natural ice can be, but, unfortunately, as the demand has increased, ice has been placed on the market, both here and in the United States, which has been harvested from less commendable sources. As a result, inquiries have been made and a demand has sprung up for hygienic ice, or, in other words, artificial ice made from pure distilled water. It is due to the fact that this demand can be readily and economically met that a great impetus has recently been given to the establishment of ice factories.

The early manufacturers of ice were quite content to turn out an article which was cold, and nothing else. No special care was taken to make the ice either pure or clear, and the cans were filled direct from the public water supply, or with water pumped from a well sunk on the premises. Given a pure water supply, it is quite possible to make marketable ice direct from the water by agitating it during freezing, so as to expel the air, and there have been many inventions for this purpose. They are, however, very nearly all open to some objection or other, the majority being cumbersome in themselves, unreliable in their performance, and very much in the way when the cans have to be taken out of the tanks. Many of them also have to be carefully watched to prevent their being frozen in.

Clear ice can be made from ordinary water by means of the plate system, and more or less clear ice by means of the cell system. With this latter it is found that the lower the temperature, the more cloudy the ice, and *vice versâ*, and both systems are open to objection on account of the length of time that the blocks take to freeze, and consequently the size of the plant which is necessary for a given output.

All the troubles relative to these systems can be avoided by distilling and de-aërating the water before it is put into the freezing-cans. This process has met with great success, and its objects may be summarized as follows:—

1. The de-aëration of the water.
2. The purification of the water.
3. The making, with absolute safety from contamination, of clear pure ice from water obtained from a well sunk on the site of the factory, so saving the expense of buying the water from the public supply.

There is no doubt that the public prefer clear ice to semi-opaque ice, and that they will be willing to pay a higher price for it; also there is a commercial value attached to the expression "Ice made from pure distilled water," as signifying freedom from pathogenic bacteria.

The distilling plants used in ice factories may be divided into two large classes:—

1. Those using the exhaust' steam from the engines, and
2. Those using live steam in multiple effect evaporators.

**Exhaust Steam Distilling Plant.**—The distilling plant which is most commonly used in ice factories is shown diagrammatically in fig. 77. In this arrangement the distilled water is obtained by condensing and purifying the exhaust steam from the engines, live steam being added to make up any deficiency. In the diagram, the exhaust steam from the refrigerating engines and other machinery comes up the pipe *A*, which is fitted with an automatic relief valve, loaded so as to give the steam a back pressure of about 5 lbs. on the square inch, and also for use in case it is necessary for any reason to exhaust direct to the atmosphere. When the distilling apparatus is in use this valve is closed, and the exhaust steam is passed through the steam-scrubber *B*, where it is robbed of its grosser impurities, such as lubricating oil, etc., and enters the feed-heater *C*, which may be of some well-known type, such as the Berryman. From this it passes into the surface condenser *D*. As there is no vacuum, no air-pump is required. The condensing water is the waste from the condensers of the refrigerating agent. As soon as the steam is condensed it runs off into the reboiler *E*, where it is violently reboiled by a steam-coil. Any air it may have absorbed is driven off, and the grease which has been passed by the steam-scrubber, and other solid matter, rises to the surface and is skimmed off through the pipe $E^1$. Preferably the skimmer should be a separate vessel. The water then flows into the cooler *F*, where it is cooled. It then passes through the filters of

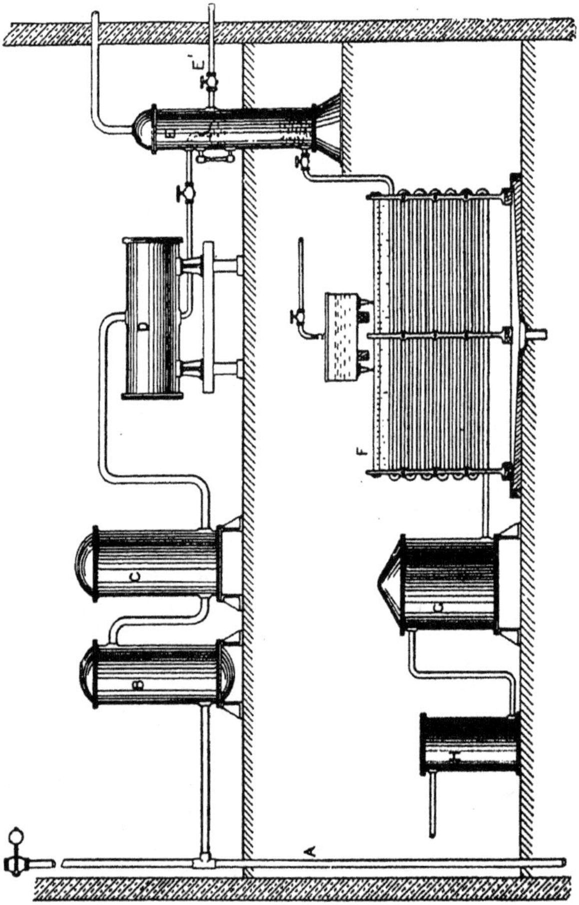


FIG. 77

charcoal or other substance *G* and *H*, and flows away to the storage tank which is kept carefully covered. In this tank it is cooled by the cold gas returning from the expansion coils, after which it is run through a sponge filter into the ice-cans.

The steam-scrubber or oil-separator, as well as the feed-water heater, may be of any of the standard types, and do not need special mention. The steam condenser may be of the ordinary surface condenser type or of the Baudelot type; as a matter of fact it is usually of the latter. In all cases the exhaust pipe should be fitted with a lightly loaded relief valve, so that, should a stoppage take place in the system, there is an alternative escape to the atmosphere. The steam enters the condenser at the top, and the condensed water runs away at the bottom. About 50 feet of 1-inch pipe per ton ice-making capacity is a usual allowance for condensing surface. The water used for condensing the steam is invariably the waste water from the ammonia condenser, which passes over the steam condenser on its way to the sewer.

**The Skimmer.**—The condensed steam enters the skimmer by a pipe at the bottom which bends upwards in order to allow any oil which has escaped the oil-filter and come through the condenser to rise to the surface and escape. The skimmer is usually a cylindrical vessel of galvanized iron with a dished bottom and wash-out valve. It has a skimming-pan at the top, and is usually, though not always, fitted with a steam-coil, which should be of galvanized iron, fed with live steam from the boiler. This coil causes violent boiling, and the impurities overflow the edge of the skimming-pan and flow away through the waste pipe.

**The Reboiler.**—The water leaves the skimmer[1] by the pipe, which is situated rather near the bottom of the vessel, and passes to the reboiler, entering it near the

[1] In some plants the reboiler comes first, then the skimmer.

top. The two vessels are usually placed side by side, and are connected together by a straight pipe. The main purpose of the reboiler is to drive off any air that may be held in the water. It is fitted with a steam-coil similar to that in the skimmer, but with a greater surface, as ebullition is required to be more rapid. In both these vessels the live steam from this coil should be condensed and drained away through a steam trap, and not allowed to escape into the water.

The reboiler is made with a dished bottom and wash-out valve and is fitted with an overflow pipe. The supply of water to the storage tank is regulated from it. In some designs the regulation is done by a float on the valve spindle, but this has not been found to work well, as the violent agitation of the water when boiling causes the float to dance and makes the valve work unevenly. If a float is used, the best results have been obtained from those of the inverted bucket type with an open bottom. The copper cylindrical air-tight floats are liable to give out and leak. The better method, however, is to put the float in a small tank adjoining the reboiler and in connection with it. Here it will be in still water and will be able to work the valve regularly. The reboiler should be fitted with a small skimming-funnel, whose height is regulated by the position of the float. A constant skimming should go on from this vessel, but the float should be so adjusted that very little water is skimmed off at a time. All surplus water should be got rid of at the reboiler and never at the distilled water storage tank, which should be as air-tight as possible, and fitted with a ball valve for regulating the supply. In some plants the skimmer is omitted, and the reboiler, which varies largely in design, is relied on to do all the skimming.

**Cooler.**—The water, now largely purified and boiling hot, leaves the reboiler by a pipe near the bottom and passes to the cooler, where it is cooled down before being passed through the filters. This cooler is of the ordinary Baudelot type, and is usually made of 1-inch galvanized iron pipe so as to get a quick heat exchange. Its size varies with different designs and with the temperature of the cooling water, from 40 to 50 feet-run of 1-inch pipe to the ton ice-making being a common rule. The steam condenser, skimmer, and reboiler, should be situated at the high level of the system, and the cooler and filters at the low level. These latter should, if possible, be below the level of the forecooler storage tank. If the ball valve in the storage tank is tight, the cooler and filters, and the pipe leading to them from the reboiler, are, by this means, always kept charged with water up to the level of the reboiler overflow. This effectually prevents the water from being drawn away and the cooler and filters from becoming charged with air.

**Filters.**—The function of the filters is to cleanse the water of any impurities that may remain in it after it has passed the skimmer and reboiler. They vary in design; a height of from 3 to 3½ times the diameter is an ordinary proportion. The filter consists of a cylindrical shell of steel, as shown in elevation in fig. 78.[1] The water enters at the bottom and leaves at the top. There is a perforated plate at top and bottom to hold the filtering material in place, but beyond this there are no internal arrangements. There are usually two filters arranged in series; generally the filtering material in both is charcoal, but in some plants the first is a silica filter. In that case it is usually rather larger than the other, and is

[1] Messrs the Frick Company, Waynesboro, Pa., U.S.A.

designed for about 50 cubic inches capacity to the ton ice-making. Sometimes sand is used as a filtering material.

The charcoal filter does not as a rule have a capacity of more than about 40 cubic inches to the ton ice-making

FIG. 78.

capacity of the plant. Sometimes all the filters are arranged in duplicate and cross-connected, but this is hardly necessary where a good skimmer is carefully used. In any case the filter should be provided with valves on the supply and delivery pipes, and with a valve on the top cover. This should be connected to the fresh water supply, the object being to force fresh water through the filter in a direction opposite to the normal flow of the condensed water. This fresh water escapes by the purge valve at the bottom, and is very useful for cleaning purposes, considerably prolonging the life of the filtering material.

**Forecooling Tank and Storage Reservoir.**—The storage reservoir consists of an insulated wood or iron tank with a tight-fitting cover, and having a capacity of somewhere about 18 to 20 gallons to the ton ice-making. The tank should be fitted with a gauge glass and thermometer in order that the level and temperature of the water can be seen at a glance, and with a purge valve so that it may be cleaned out at intervals. The supply of distilled water to it should be regulated by a ball valve, and the tank should be of such a size that this valve is nearly always running. It is not advisable to keep more water stored in the tank than can be helped, or it will absorb air and other gases, and produce cloudy ice. The cooling of the water in the tank is usually effected by coils through which the suction gas passes on its way to the compressor. In some plants, however, this water is cooled by an independent direct expansion coil fed from the liquid-receiver. It is very desirable to cool down the water in the forecooler to as near freezing-point as possible, for not only is it less likely to absorb air and gases at this temperature, but it saves capacity in the ice-tank.

**Sponge Filter.**—After leaving the storage tank the water is finally filtered by being passed through a sponge filter. This should be long and narrow, and, to ensure good results, should have a length of from 6 to 8 times its diameter. The capacity varies, but the larger the better. A capacity of 100 cubic inches to the ton ice-making has been adopted in some cases. The sponges should be close grade Turkey, and tightly packed into the filter. A spare set should be kept, and they should be changed once every twenty-four hours. After being taken out of the filter, the sponges should be carefully washed and kept in a can of clean water until they are wanted again.

**Filler.**—From the sponge filter the water is conveyed by flexible hose to the ice-cans, and is filled into them by

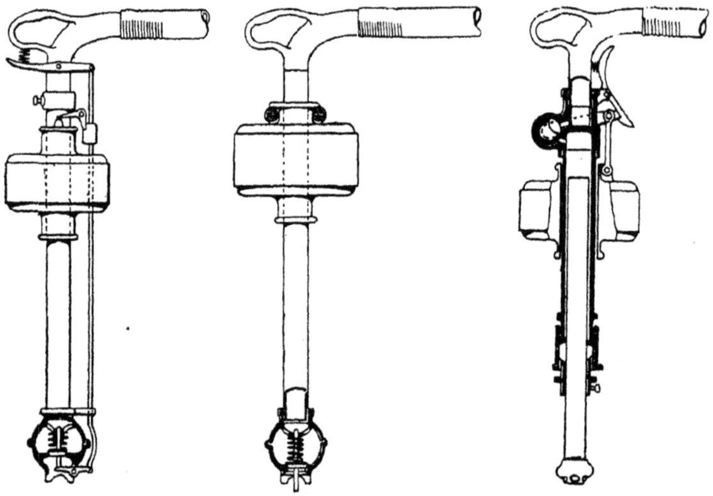

FIG. 79.

a proper can-filler. There are several good patents on the market, and, though they differ in detail, the principle is largely the same. Fig 79[1] illustrates their designs. The filler is put into the can, and, as it touches the bottom, a

[1] *Machinery for Refrigeration*, Norman Selfe, page 250.

valve opens and allows the distilled water to run into the can without splash or fall. When the can is filled to the proper level, the water supply is automatically cut off by the float, the filler is removed, and the can covered in.

**Exhaust Steam Triple Effect.**—It is a well-established fact that the steam required to work a good compound engine will not be sufficient, when condensed, to provide all the distilled water necessary for ice-making purposes. Where such engines are in use, it is, or has been, the almost universal custom, brought about by necessity, to carry live steam from the boilers direct to the steam condensers in order to supply the shortage. It will be readily seen that this method is an exceedingly wasteful one, for steam is raised and condensed without having served any useful purpose, and no use whatever is made of the latent heat which goes away in the condensing water.

In considering this question of distilling water, it must be realized that at atmospheric pressure it takes 966 B.T.U. to turn 1 lb. of water at 212° F. into 1 lb. of steam at the same temperature, while afterwards it only requires the addition of 126 B.T.U. to raise that pound of steam to 100-lb. pressure on the square inch. The 126 B.T.U. are used up in the steam-engine, but the 966 B.T.U. remain in the steam and have to be taken out in the condenser before the steam can be turned into water. This latent heat requires coal, and so costs money to make, and it is only right that it should be made use of if possible. Fortunately the means of doing this are ready to hand, and, by the use of a double or triple effect, it is possible to use the latent heat and to obtain the boiling and evaporation of a larger quantity of water than is supplied to the apparatus in the shape of steam. A diagram of a triple effect of the most simple kind, as arranged for distilling water, is shown in fig. 80.

The principle on which it works is the physical law that, in condensing to a liquid, the sensible heat of vapour is

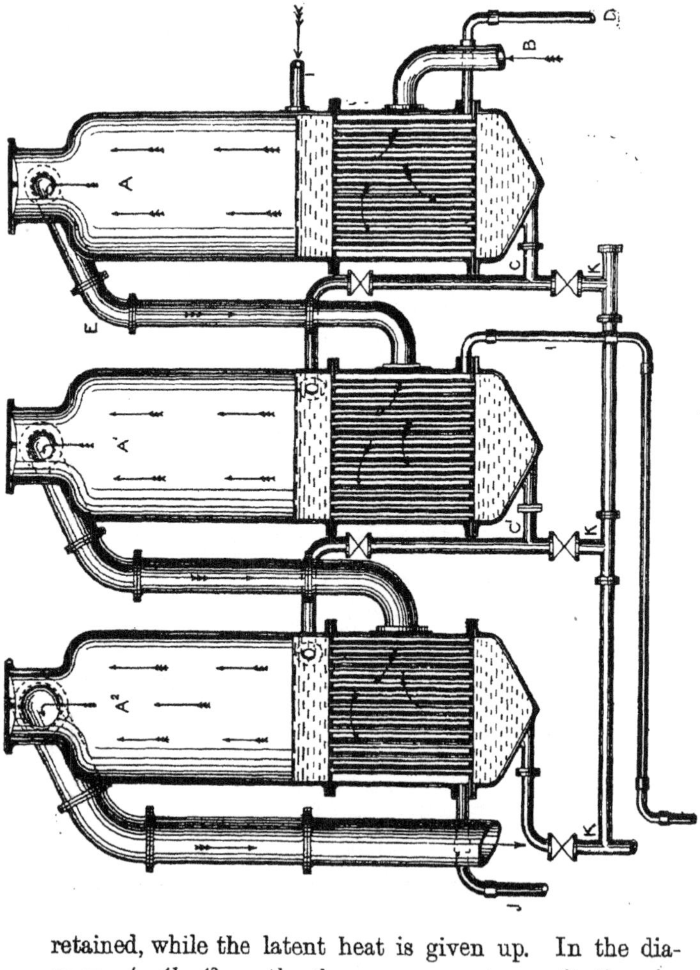

FIG. 80.

retained, while the latent heat is given up. In the diagram, *A*, $A^1$, $A^2$ are the three pans or pots constituting the effect. Each of these pans is divided into two parts, an upper and a lower. These two parts are connected together with a number of tubes; the space round these tubes is filled, in the case of *A*, with exhaust steam from the engines which enters through the pipe *B*, and in the case of $A^1$ and $A^2$ with vapour from the preceding pan. The water to be evaporated enters through the pipe *L* and circulates inside the tubes and above and below the tube plates. The top of the pan *A* is connected to the steam space of the pan $A^1$, and the top of that to the steam space of $A^2$, the top of $A^2$ being connected to the surface condenser. The lower portion of the liquor space in *A* is connected by the pipe *C* to the upper portion of $A^1$, and the lower portion of $A^1$ by the pipe $C^1$ to the upper portion of $A^2$. It will thus be seen that, as soon as the vacuum is started in the condenser, it will be felt in decreasing power throughout the apparatus. Now, if a vacuum of 4 inches to 5 inches, corresponding to a temperature of 203° F., is maintained in *A*, the water in it will readily be evaporated by the exhaust steam from the engine at a temperature of 223°. The vapour from this pan will likewise readily boil the liquor in $A^1$ under a vacuum of 14 inches corresponding to a temperature of 182° F., and the vapour from that will readily boil the liquor in $A^2$ under a vacuum of 25 inches corresponding to a temperature of 125° F.

When the apparatus is at work the exhaust steam enters through the pipe *B* and boils the first effect *A*, the condensation being withdrawn through the pipe *D* and returned to the boiler. The vapour from *A* passes through the pipe *E* and boils the second pan $A^1$. The vapour given off by the boiling water in $A^1$ boils the water in $A^2$, the vapour from which is condensed in the surface condenser.

The water passes from pan to pan on account of the difference of vacuum, and the distilled water is taken by the pipes *I* and *J* attached to the subsidiary pumps. Each pan can, when idle, be emptied by means of the pipes and valves *K*. In practice the tubes are generally laid horizontally with the steam or vapour inside, and about 6 lbs. of water evaporated per square foot of heating surface per hour is very satisfactory.

The distilled water is delivered from the pumps attached to *I* and *J*, and from the air-pump of the condenser, into a tank placed in some convenient position, from which it is taken and delivered to the ice-cans.

The advantages of this apparatus are:—

1. That by making use of the latent heat, for every pound of steam used in the first effect it is possible to obtain from 2 to 2½ lbs. of pure distilled water, instead of under 1 lb., as in the apparatus described in fig. 77.
2. The steam from the engine, with all its impurities, is, after having the oil removed, returned to the boiler as feed-water, and is not used for ice-making.

The disadvantages are:—

1. The back pressure on the engine causes water in the cylinders, and there is an absence of the economy obtained from working condensing.
2. The large amount of heating surface which has to be provided, and therefore the increased first cost.
3. The trouble of having to draw off the product of the various effects through vacuum pumps, and the expense of keeping these in order.

**Live Steam Distilling Plant.**—These disadvantages can be overcome in a somewhat similar apparatus which will give an equal quantity of water per pound of primary steam, and in connection with which the refrigerating and other engines are worked with a vacuum in their cylinders, their exhaust steam being condensed in a surface condenser in the usual way.

With this apparatus the first effect of the evaporator is quite independent of the engine, and is heated with live steam from the boilers reduced to a pressure of about 25 lbs. to the square inch above atmosphere. The last effect exhausts into a condenser at atmospheric pressure. The water to be evaporated passes through an exhaust-steam heater placed on the exhaust pipe of the engine and between the engines and the condenser, and so reaches the first effect in a very hot state.

The advantages of this apparatus are:—

1. No back pressure on the engine, and, consequently, an economy in its steam consumption.
2. The small amount of heating surface necessary.
3. The small space occupied.
4. The low first cost.
5. The convenience of being able to draw off the product of the effects by gravity and without an air-pump.
6. The fact that the distilling apparatus is entirely independent of the rest of the plant.

With regard to No. 2—considering that the difference in total heat between exhaust steam at 5 lbs. and live steam at 25 lbs. is very small, there would seem theoretically to be no reason why the heating surface should be less with live steam than with exhaust, but it has been found in practice that the little extra pressure makes a considerable difference, reducing the necessary heating surface considerably.

In order to work a triple effect, it is necessary to have a certain range of pressure. Thus, if the exhaust steam is at a pressure of 5 lbs., then the vacuum in the condenser must be equal to 12 lbs.; but if the exhaust steam is at a pressure of 25 lbs., then there need be no vacuum in the condenser. The process of interchange of heat between the various effects is exactly the same as already described, and the latent heat is made use of in exactly the same way.

Unfortunately, as will be shown later on, though the live-steam effect is an exceedingly valuable piece of apparatus, it cannot compete in economy with the same apparatus using exhaust steam. It is of great value, however, in cases where only a small portion of the power of the engines is used in making ice, or where only a small portion of the ice produced is made from distilled water, as then the main engines can be worked condensing, and economy secured in this way.

A live-steam triple effect is illustrated in fig. 81.[1] It works on the same principle as the sextuple effect about to be described. The apparatus as used for exhaust steam is exactly similar in principle, but has an air-pump, and, owing to the greater heating surface required, is larger for a given output than the one illustrated.

It is, of course, quite possible to heat the primary coils with live steam and at the same time have a high vacuum in the condenser. In a double or triple effect this is of no material advantage, but it is the only possible method of economically using the sextuple effect.

A multiple effect apparatus, manufactured by the same

[1] Messrs the Mirrlees, Watson Company, Ltd., Glasgow.

Fig. 81.

makers, is illustrated in fig. 82. Its characteristic feature consists in film evaporation.

The principle will be better understood by reference to fig. 83, which represents diagrammatically a section through one of the effects. The action is as follows:—

The water to be distilled, and to be afterwards used for making ice, is admitted to the inside of the tubes through the pipes *A A*. Live steam, or the vapour from the preceding effect, is admitted through the opening B and plays round the tubes. As soon as the liquid, which is already nearly boiling, enters the tubes it begins to boil, and, feeling the influence of the vacuum, it follows the course indicated by the arrows and continues to absorb heat, until, finally becoming a mass of spray, it rushes through the tubes and impinges on the metal plates, which cause the water to separate from the vapour; the former, falling to the bottom, runs off through the opening *C*, to be further evaporated in the next effect; the latter passes away through the opening *D*, to expend its latent heat in boiling in a greater degree of vacuum in the next effect the water from which it has just been evaporated. The condensed steam or vapour from the effect is withdrawn through the pipe *E*. This process continues throughout the six effects.

The feed-water is heated on the "step-up" principle. In the first instance, if the water is clean enough, it is taken warm from the circulating water of the condenser. It is then exposed to the vapour of the last effect while this is passing to the condenser, and then to the vapour of the several effects constituting the apparatus, until, just before entering the first effect, it is brought to the boiling-point of that effect by the live steam heating it.

A sextuple effect, with a pressure of 40 lbs. of steam in

Fig. 82.

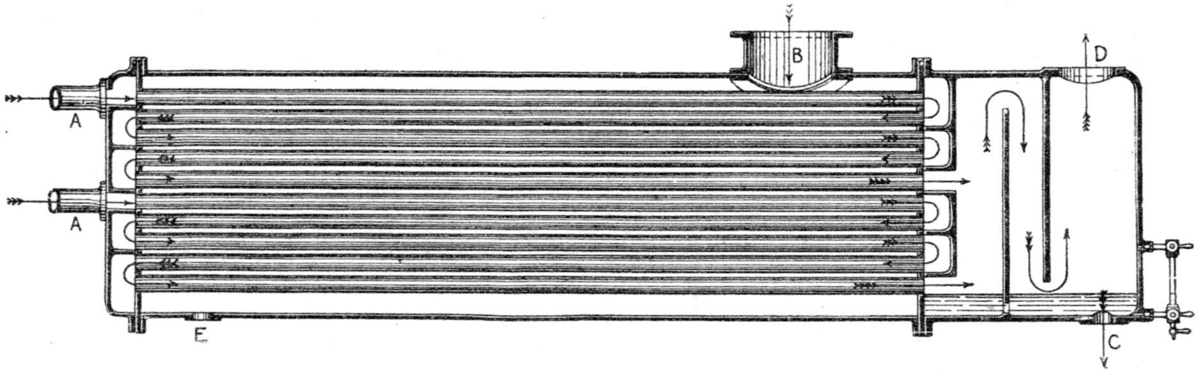


Fig. 83.

its first effect and a vacuum of 27 inches in its condenser, is capable of producing $4\frac{1}{2}$ lbs. of distilled water for every pound of live steam used in the first effect, but, owing to live steam being a necessity, its application is limited in ice factories where ample exhaust steam is usually available. The production of the various sizes is approximately as follows, the primary steam being in all cases returned to the boiler as feed-water :—

| | | | | | | | |
|---|---|---|---|---|---|---|---|
| Double effect | | $1\frac{1}{2}$ lbs. | distilled | water | per lb. | of | steam. |
| Triple | ,, | $2\frac{1}{4}$ ,, | ,, | ,, | ,, | | ,, |
| Quadruple | ,, | 3 ,, | ,, | ,, | ,, | | ,, |
| Quintuple | ,, | $3\frac{3}{4}$ ,, | ,, | ,, | ,, | | ,, |
| Sextuple | ,, | $4\frac{1}{2}$ ,, | ,, | ,, | ,, | | ,, |

It must be remembered that water produced by this means is ready for making clear pure ice without any further treatment, except, perhaps, that of passing it through a filter. Sometimes it is reboiled to rid it of any air or permanent gas it may still have in it, but this is not usual. In any case the net result is that with a triple effect $2\frac{1}{4}$ lbs., and a sextuple effect $4\frac{1}{2}$ lbs., of pure distilled water, free from all pathological microbes and ready for making pure clear ice, are obtained for every pound of boiler-steam fed into the apparatus, or (in the latter case, and assuming the evaporative efficiency of the boiler and coal to be only 8 lbs. of water to 1 lb. of coal) 36 lbs. of water per lb. of coal.

**Working Cost.** — Having now described the various apparatus that are or might be used in producing pure distilled water for ice-making, it will be interesting to go into the question of the cost of the plant for producing it, for it is useless to recommend an article unless it can be clearly shown that money or its equivalent can be saved by its use.

In most towns pure water can be bought at the rate of 6d. per 1000 gallons, or 1¼d. per ton, so that, when ice is only wanted to be cold, and there is no objection to its being opaque, it obviously will not pay to distil the water. But, granting that it has been decided for various reasons that pure hygienic ice should be made, the only question to be solved is:—What is the cheapest and best method of producing it, taking the coal first and considering labour, interest, and depreciation afterwards?

Four apparatus have been described:—

No. 1. That shown in fig. 77.

No. 2. That shown in fig. 80.

No. 3. Similar in principle to No. 2, but working with live steam.

No 4. The sextuple effect evaporator.

Taking a hypothetical case of a plant making 50 tons of ice per twenty-four hours, and applying each method to it, allowing throughout that 100 I.H.P. is required:—

### With No. 1 Apparatus.

There is a 100-I.H.P. engine working under a back pressure of 5 lbs. per square inch; 35 lbs. of steam per I.H.P. hour will under these conditions be a fair estimate of its consumption. This will represent 37·5 tons of steam per twenty-four hours as the engine exhaust.

Thus there is:—

| | Tons. |
|---|---|
| Exhaust from engine, . . . . . . | 37·5 |
| Balance to be made up by condensing live steam, | 12·5 |
| [1] Total, . . . . | 50 tons |

[1] In practice more than 50 tons will have to be evaporated to make up for loss in skimming, etc., further increasing the cost.

evaporated, which at 8 lbs. = 6·25 tons of coal, which, taken at 10s., would make the total steam used by the plant cost £3, 2s. 6d.

### With No. 2 Apparatus.

The engine is still working under a back pressure of 5 lbs., and using, say, 35 lbs. of steam per I.H.P. hour, or 37·5 tons per twenty-four hours. This steam is being exhausted into the first effect of a 2½ to 1 triple effect, so that the 37·5 tons would produce, say, 94 tons of distilled water, or nearly twice as much as is wanted. Consequently the evaporator will only have to work thirteen hours per day, the engines exhausting to atmosphere or condensing the rest of the time, and so working at a better efficiency, or an evaporator of half the size and working for the twenty-four hours might be used.

Thus there is:—

| | | | | Tons. |
|---|---|---|---|---|
| The engine working for | 13 hours | at | 35 lbs. = | 20·3 |
| " " | 11 | " | 30 lbs. = | 14·7 |
| Total steam used by engine, | | | . . | 35·0 tons. |

The 20 tons used in the thirteen hours will produce in the triple effect 50 tons of distilled water.

So that the total steam used by the plant is 35 tons, requiring 4·37 tons of coal, and costing £2, 4s. per twenty-four hours.

### With No. 3 Apparatus.

The 100-I.H.P. engine is working to a condenser, and so is only using 25 lbs. of steam per I.H.P. hour, or 26·78 tons per twenty-four hours.

But the steam for the evaporator is in addition to this,

and 50 tons have to be evaporated in the apparatus at 2½ to 1, which will require 20 tons of live steam.

Thus there is:—

| | Tons. |
|---|---|
| Steam used in the engine, say, . . . . | 26·5 |
| Steam used in evaporator, say, . . . . | 20 |
| Total steam used, . . . | 46·5 tons, |

requiring 5·8 tons of coal, and costing £2, 18s. per twenty-four hours.

WITH No. 4 APPARATUS.

The 100-I.H.P. engine is worked to a condenser, and is only using 25 lbs. of steam per I.H.P. hour, or 26·78 tons per twenty-four hours.

But the steam for the sextuple effect is in addition to this, and 50 tons have to be evaporated in the apparatus at 4½ to 1, which will require 11 tons of live steam.

Thus there is:—

| | Tons. |
|---|---|
| Steam used for engine, say, . . . . | 26·5 |
| Steam used for evaporator, . . . . | 11 |
| Total steam used, . . . | 37·5 tons, |

or, say, 4·7 tons of coal, costing £2, 8s. per twenty-four hours.

To summarize:—

| | £ | s. | d. |
|---|---|---|---|
| No. 1 plant costs in coal per 24 hours, | 3 | 2 | 6 |
| No. 2 " " " | 2 | 4 | 0 |
| No. 3 " " " | 2 | 18 | 0 |
| No. 4 " " " | 2 | 8 | 0 |

Here No. 1 may be left, for, in dealing with this item, it is necessary to generalize largely, as no reliable data exist as to its cost of installation, the exact practice and the apparatus employed varying largely with the will of the designer of each installation. It must be remembered, however, that though the first cost of the plant is probably low, there is a constant charge for filtering material and for labour. The attention required is much greater than in the other systems, while any attempt to economize in these two items is liable to give the ice a bad taste, and so spoil a freezing. The system, however, has been largely used in the United States, and, although the risk of getting impure ice is greater than with the other systems, there is no doubt that, where care and attention have been given to the cleanliness of the various vessels, its use has been attended with success.

No. 3 system has already been commented upon, and its advantages and limitations pointed out.

With regard to No. 2, the most serious disadvantages are the first cost and the back pressure on the engines, and consequent danger of getting water in the cylinders.

Where exhaust steam can be obtained in ample quantities, and there is no objection to working the engines with a back pressure, its economy is most marked.

Taking, for instance, the hypothetical example before referred to, the cost in coal by this system was £2, 4s. per twenty-four hours, or, assuming the plant to be working full power for only six months in the year, say £401. The first cost of the apparatus would be about £1750; allowing 10 per cent. for interest and depreciation, the total annual cost would be £576.

System No. 1 costs £3, 2s. 6d. per twenty-four hours, or, for the six months, £570.

Assuming that interest and depreciation, filtering material, and extra labour would cost for system No. 1

£200, the total annual cost would be, say, £770, showing a saving for No. 2 of £194 per annum.

In the above the triple effect is only working thirteen hours out of the twenty-four; if a smaller size were installed and allowed to work the twenty-four hours round, it is easy to see that, as the first cost would be less, the saving would be greater.

It must, however, be borne in mind that in a modern ice factory of any size the refrigerating machines are frequently required to deal, not only with the ice-tanks, but with cold storage as well, and there are also electric lighting and power machinery, pumping machinery, and possibly air-compressors and other auxiliary machinery. In fact, the power used in ice-making is but a proportion of the whole power generated. There are obvious difficulties in splitting the exhaust-steam supply and working some of the engines under a back pressure to supply the distiller, and some condensing, and as it is out of the question, from an economical point of view, to run the whole plant under a back pressure for the sake of utilizing a small portion of the exhaust steam in the distilling apparatus, it may be desirable to exhaust the whole of the engines to a surface condenser and have the distilling plant entirely separate and self-contained. In this case a live-steam triple effect or a sextuple effect would probably be the best plant to install.

For instance: The first cost of a 50-ton sextuple effect working with live steam is about £1800. 10 per cent. for interest and depreciation would represent £180 per annum. It costs in steam £2, 8s. per twenty-four hours, or for six months £438, making a total of £618, which is only £42 more than No. 2.

**The Lillie Evaporator.**—The Lillie evaporator, which is

much used in the United States for producing distilled water, is also worked on the Film Evaporation principle. A section of one of the effects is shown in fig. 84.[1] The

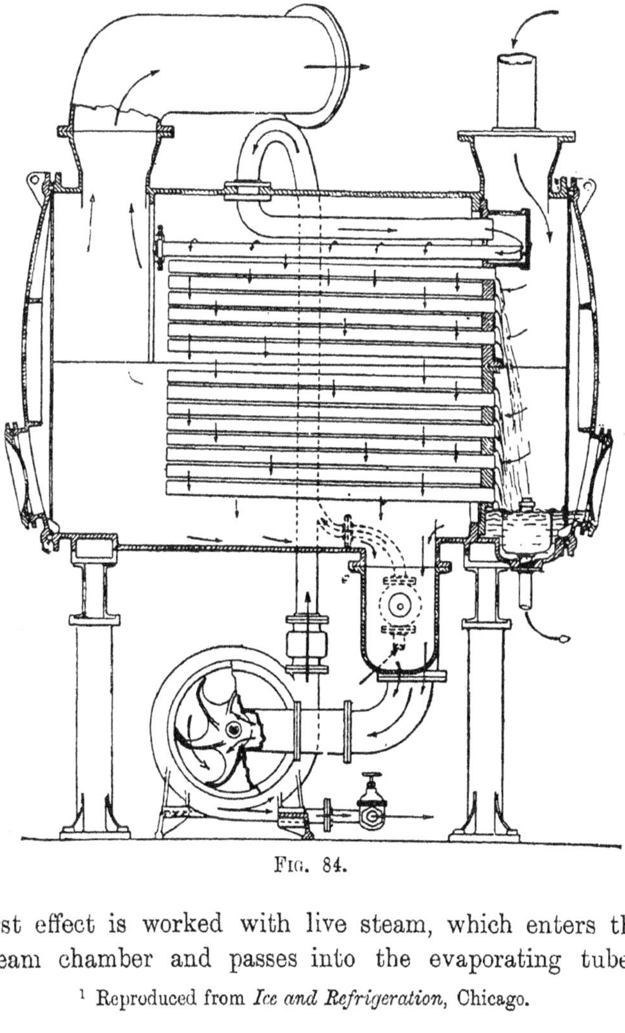

Fig. 84.

first effect is worked with live steam, which enters the steam chamber and passes into the evaporating tubes.

[1] Reproduced from *Ice and Refrigeration*, Chicago.

The water to be evaporated is pumped by the centrifugal pump into the manifold at the top of the evaporating tubes. From this it flows into the slotted distributing pipes, one of which is placed over each row of evaporating tubes. There is at once an interchange of heat between the steam in the tubes and the water flowing over them, the one being condensed and the other evaporated. The vapour from the evaporated water passes to the steam space of the next effect, where, under a lower pressure, it evaporates more water. A full account of this evaporator will be found in the Journal from which the figure is taken. The net result of the test of the apparatus was an evaporation of nearly 14 lbs. of distilled water per lb. of coal, or not quite 2 lbs. of water per lb. of steam. With this apparatus the following method of cleaning the tubes has been tried, though with what success is not known:—

A solution of carbonate of soda is run through the effects, which are heated at the same time; this solution, $Na_2CO_3$, acts on the insoluble scale, which is largely sulphate of lime, and soluble sulphate of soda and carbonate of lime are formed, the last being soluble in hydrochloric acid. The $NaSO_4$ passes off in the water. A dilute solution of hydrochloric acid is then run through the heaters and effects. The remaining carbonate of lime reacts with the acid, forming soluble calcic chloride, carbon dioxide, and water. All soluble salts pass off with the solution.

### Defects in Ice due to the Use of Distilled Water.

When distilled water is produced from live steam as in the evaporators just described, it is fairly pure and quite dead and airless. Filtering is not generally required, but a charcoal filter and a sponge filter are always useful adjuncts, particularly as it is one of the peculiar properties of the former to absorb any bad smell or odour that the water may have. When, however, ice is made from distilled water produced by condensing the exhaust steam of the engine, the way is paved for numerous defects in the ice. The most prominent of these are as follows:—

**Red Core.**—This is a reddish tinge which shows itself in the centre of the blocks of ice, and can generally be traced to rust in the distilling plant. It usually reveals itself during the second season after the first winter's rest. To avoid this rust the distilling plant should be of galvanized iron throughout. Sometimes the inside of the filters and tanks are painted at the beginning and end of each season. It is advisable to keep the distilling plant full of distilled water during the winter. Red core can frequently be got rid of if the forecooler chills the water to from 36 to 33° F. before it is passed through the sponge filter. Sometimes the water itself contains iron oxides—this can be ascertained by analysis—which are much the most troublesome to deal with. When hard town water, or water derived from a well in the chalk, is used for ice-making, it will be found to contain, not only carbonate of lime, but carbonic acid, which, being a gas, passes over with the steam. As a matter of fact, distilled water nearly always contains a small proportion of carbonic acid, which, in contact with the iron pipes of the plant, forms two kinds of iron oxide—one the ordinary ferric oxide or iron rust, which is caught by the filter; and the other a solution of ferrous oxide, which is not caught, and which passes to the ice-cans.

When the water is being frozen, what little air it may still contain is brought into contact with the oxide, which is turned into rust and causes the red core. No amount of filtering will get rid of this oxide. The only satisfactory way is to cure the evil at the start and soften the water before it enters the boilers by means of suitable and cheap chemical reagents. These get rid of the carbonate of lime, iron salts, and magnesia, which are the chief scale-forming impurities, and not only keep the ice clear, but considerably prolong the life and efficiency of the boiler. Water-softening has been dealt with in Chapter VII.

There is another reason why it is important to eliminate the carbonic acid, for this substance in contact with blackbone will produce carbonate of lime, giving a white core to the ice.

Red ice is not always due to rust, but to a peculiar deposit from the lubricating oil used in the cylinder. This should always be caught by the skimmer, and seldom reveals itself unless the plant is being forced. If the red deposit is due to oil, it will float in warm water and sink in cold, and can be readily tested in this way. When oil is appearing in the distilled water, the oil-separator is not doing its duty, and the chances are that the steam condenser will be found coated with oil. If so, it must be blown out with live steam [1] until it is clean. A flannel bag placed over the delivery into the forecooler tank has been recommended, but this is hardly necessary where a good sponge filter is in use.

**White Core.**—The usual cause of white core or cloudy ice is the presence of air in the water, but it can be caused in other ways, and is sometimes due to carbonate of lime and magnesia, which has been carried over with the steam, or which has been formed by carbonic acid coming in contact with the blackbone in the filters. It

[1] See method of cleaning surface condensers, page 182.

is also due to the priming of the boiler and to impurities derived from overworked filters. As before stated, most of these troubles can be got rid of if the water is fore-cooled to 33 or 34° F.

**Charcoal for Filters.**—Probably the best insurance against bad ice is the use of good charcoal. It is well known that this substance will chemically absorb odours and purify the water, as well as mechanically rid the water of any impurities. Almost any hard wood charcoal will do for filters. In the United States sugar maple is preferred, but beach charcoal is equally good. Charcoal should be renewed from time to time, as it loses its properties of absorption, and these cannot be restored to it.

**Starting up a Distilling Plant.**—When a distilling plant is first put down, and at the beginning of every season, it should be thoroughly cleaned. In the first case oil, red lead, and dirt will have been left in the pipes and coils by the workmen, and, if not got rid of, will spoil the ice. These impurities can generally be displaced by blowing out the apparatus with hot water and live steam; but there is another source of trouble due to the zinc oxides from the galvanized pipes becoming dissolved in the distilled water. As a precaution against this, it is recommended that a plant before being started should be filled with a strong solution of soda ash. After this has stood for several hours, it should be blown out with live steam or live steam and hot water. Of course, before this is done the charcoal must be carefully removed from the filter, for, if the soda once gets into that, it means new charcoal.

**Blackbone and Blood Charcoal.**—If, after taking all the above precautions, the ice is still unsatisfactory from either bad taste, smell, or colour, a cure can generally be effected by mixing finely powdered blackbone or bone charcoal with the distilled water. This is a powerful absorbent, and a pound or two of blackbone will be sufficient to withdraw the taste or smell from a ton of water. Blood charcoal, which is a bye-product in the manufacture of prussiate of potash, will do equally well, but is rather more expensive.

# CHAPTER XI.

## ICE-MAKING.

ONE of the most important uses to which mechanical refrigeration has been put is the artificial freezing of water to produce ice.

Water freezes at a temperature of 32° F., and in freezing changes its chemical constitution. In order to effect the change, the latent heat must be withdrawn from the water. The latent heat of ice is 142·4 British Thermal Units—that is to say, in order to convert 1 lb. of water at 32° F. into 1 lb. of ice at 32° F., 142·4 B.T.U. must be withdrawn by the refrigerating machine. To this latent heat must, of course, be added the sensible heat of the water, and, assuming that when freezing operations were started the water was at a temperature of 60° F., $60-32+142·4$, or 170·4 heat units per lb., would have to be abstracted before it would be completely frozen. After freezing is complete, the block of ice is reduced to a temperature of about 20° F. As the specific heat of ice is 0·5, $32-20=12$, $12 \times ·5=6$ B.T.U. have still to be extracted. This makes a total of 176·4 B.T.U. per lb.

**Water for Ice-making**—Practically any water that is fit to drink will make ice fit to use, but if the ice is to be clear the air must be eliminated. The same remark applies to the carbonates of lime or magnesia or to other mineral matters that nearly all water contains in solution. Agitation, however, will not get rid of these, and they remain in the water making the ice cloudy. Following naturally from this, soft water would be the most likely to turn out clear ice, and where the natural supply is very hard it is advisable to soften the water before it is frozen. As an alternative, both carbonates and suspended matter generally, can be got rid of by vigorously boiling—not distilling—the water before it is frozen.

In the majority of plants the cans or tanks are filled direct by a hose with water taken from the public supply, and at whatever temperature it may happen to be. The resulting ice will be perfectly healthy, and care only has to be exercised when the ice is made from well water or some other private supply. The bulk of the natural ice imported into England is drawn from the Norwegian lakes, which, though fed by surface drainage, are, owing to the scarcity of population, rarely subject to contamination.

There are in general use three ways of producing ice. These are known as the can system, the plate system, and the cell system. The two former are very largely used in the United States, and the first and third in the United Kingdom.

**The Can System.**—In making ice by this system, the water to be frozen is placed in galvanized iron cans which are suspended in a tank containing refrigerated brine. The brine surrounds the cans and is kept cold by coils of pipe placed in the tank, and in which ammonia is expanding. The cans are of an oblong taper shape and about three times as deep and twice as long as they are wide, the reason being that as the water next the iron of the cans freezes first, all the heat of the water subsequently frozen in the can has to be withdrawn through this ice, and, as ice is a non-conductor of heat, it is desirable to have as small a thickness of it as possible. The water is agitated during freezing to free it from air, or distilled water free from air is used.

The brine is kept at a temperature of from 8° to 18° F., 14° being a very usual temperature.

**The Plate System.**—In the plate system, instead of immersing the cans containing water in the cold brine, the ice-tank is divided up by a number of vertical detachable partitions placed from 17 to 22 inches apart. These partitions are made either of hollow plates, through which the cold brine circulates, or of coils of pipes covered on the outside with a flat plate so as to present a smooth surface. The plates or coils are from 8 to 12 feet long and 6 feet deep, and the space between them would be, say, from 20 to 27 inches wide. The tank being filled with water and the cold brine circulated in the plates, ice begins to form gradually on each side of the plate. Freezing is continued until the ice becomes from 6 to 10 inches thick, when it is thawed off and harvested.

Owing to the ice gradually forming in this way and freezing from one side only, any air or impurities that the water may contain have a tendency to be frozen out and remain in the water left unfrozen. This is assisted by the agitation of the water, and the ice is beautifully pure and transparent. Freezing on a plate from one side of the ice cake only, takes a considerable time, and the plant is expensive both in first cost and up-keep. Sometimes the plates are put 10 or 12 inches apart, and freezing is continued until the ice on the opposite plates meets. In this case, however, the air and impurities are apt to be frozen in, the ice has a core and is not so pure or clear. When ready for harvesting, the ice is detached from the plates by turning off the supply of cold brine and turning on a supply of warm brine which thaws off the ice, enabling it to be withdrawn.

**The Cell System.**—The cell system somewhat resembles the plate system, in that the water to be frozen is placed in cells which are made with hollow iron walls in which cold brine is kept circulating. The tank containing the water to be frozen is made with a double bottom and generally of wood. It might measure from 8 to 12 feet square and 5 feet deep. Each of these tanks has a space at one side, in which a float is worked up and down, giving an ebb and flow motion to the water, which flows through the holes in the false bottom into the cell itself, and is thus kept constantly agitated. Each tank is divided up into a number of separate cells by long narrow hollow walls of iron, in which the brine circulates. Each brine wall is divided up into a number of horizontal divisions, so that the brine when circulating is forced to take a zigzag course, and is brought into thorough contact with each part of the cell. When the cell ice is ready for harvesting, the cold brine is turned off and warm brine is circulated, as in the plate system.

This brief *resumé* of the principal features of each system will give a general idea of the processes employed. Refrigerating machinery is common to all three systems, and only the actual ice-tanks and apparatus vary. The details of the three systems are as follows :—

## THE CAN PLANT.

The can plant may be divided into three general classes :—

1. Tanks in which the expansion coils are placed in the tank itself and between the rows of cans.

2. Tanks in which the expansion coils are placed in a separate part of the tank, the brine surrounding the cans being circulated through these coils and round the cans by a propeller.
3. Tanks in which the expansion coils in the tank itself are omitted, the brine being cooled in a separate independent cooler and allowed to gravitate into the tank. After having circulated round the cooler, it is pumped back through the cooler and again to the tank.

Can ice can be of three qualities :—

1. *Opaque Ice*, which is ordinary water frozen with all it contains.
2. *Agitated Ice*, which is the same, but with the water deprived of a large portion of its free air by agitation.
3. *Distilled Water or Crystal Ice*, in which the water before being frozen is distilled and rendered absolutely pure. The ice, having no air left in it, is quite clear and transparent.

**Opaque Ice** such as is shown in fig. 85 is much the cheapest to manufacture, and is extensively used for all sorts of cooling purposes. It is sold largely in a crushed form to the fishing-fleets for preserving the catch, to ice-cream merchants, confectioners, butchers, etc. for the various purposes connected with their trades, and to the ordinary householder for cooling the domestic refrigerator.

Generally speaking, it will melt more quickly and do less refrigeration than a similar sized cake of crystal ice. This is due to the fact that the air it contains makes it porous, and, by increasing the surface, enables the heat to get to it more rapidly.

**Agitated Ice** is more expensive to make than opaque

ice, and less so than distilled water ice, coming about one-third of the way between the two. If the agitating gear is of an efficient design and the water is pure,[1] it is quite good enough for all ordinary purposes.

**Distilled Water Ice.**—This is the highest class of ice that can be produced, and is as pure as it is possible to make ice. As it has been robbed of all its natural salts in the distilling process, it is rather flat and tasteless, and does not satisfy the thirst if taken in any quantity by itself. As, however, it seldom is taken in this way, but merely as a diluent, this is not a very material objection.

**The Ice-tank.**—The construction of the ice-tank is much the same for all three classes of can ice. It is made of iron plate, well stayed and braced. Sometimes steel is used, but iron is preferable, as it is less liable to be attacked by the brine. The top of the tank can be constructed so that the framework supporting the cans rests on the top of the tank; but this is not advisable, as it brings the brine level too near the top

[1] Freezing alone will not kill or even lessen disease germs, and if the water from which the ice is made is in any way contaminated, the ice will also be contaminated. The only way to absolutely guard against the presence of disease germs in ice is to distil the water before it is frozen. It was stated, in connection with some experiments recently carried out and reported to the Royal Society, that the germs of typhoid fever, diphtheria, anthrax, and cholera were subjected to a temperature of $-312°$ F. for seven days, and at the end of that period were found to be quite alive.

Professor Dewar, in his address to the British Association at Belfast in 1902, stated that: "Very severe cold does not destroy germs. A persistence of life in seeds has been demonstrated at the lowest temperatures, and Professor Macfadyen has shown that many varieties of micro-organisms can be exposed to the cold of liquid air for a period of six months without any appreciable loss of vitality, although at such a temperature the ordinary chemical processes of the cell must cease. The cells cannot be said in the ordinary sense to be either dead or alive—it is a new condition of living matter—a third state."

and may result in the brine overflowing. The method more usually adopted is to rivet an angle iron to the inside of the tank about 6 inches below the top. The can frames rest on this, and the wooden covers rest on the top of the tank. In the best practice the tanks are made with flanged corners, as angle-iron corners are troublesome and liable to leak. The tank should be provided with an overflow pipe, so that, if through any carelessness in filling, the cans are allowed to overflow and the tank get too full, the brine will merely run to waste and will not slop over and wet the insulation. With efficient filling apparatus, however, there is no danger of this happening.

The cans are arranged in the tank so that there is only enough space between them for the framework that supports them; the ammonia coils are (Class 1) placed longitudinally down the tank between each row of cans.

**Size of Cans.**—In England the cans are usually made in the following sizes. An allowance is made in the size of the cans for a loss of about 5 per cent. in thawing out.

| Width. | Length. | Depth. | Weight of ice. | Thickness of iron. |
|---|---|---|---|---|
| in. | in. | in. | lbs. | in. |
| 6 | 24 | 24 | 112 | $\frac{1}{16}$ |
| 6 | 30 | 30 | 168 | $\frac{1}{16}$ |
| 8 | 30 | 30 | 224 | $\frac{1}{16}$ |
| 8 | 36 | 36 | 336 | $\frac{3}{32}$ |
| 8 | 42 | 42 | 448 | $\frac{1}{8}$ |
| 12 | 42 | 42 | 672 | ... |

With a temperature of about 14° in the tank, the time

taken to freeze can ice is about half the square of its thickness—*viz.*, 8-inch ice $= 8 \times 8 = 64$, $\frac{64}{2} = 32$ hours. The time taken to freeze plate ice is about twice the square of its thickness.

The following table has been compiled by Mr Siebert, and gives the time required for freezing with different temperatures and thicknesses.

FREEZING TIMES FOR DIFFERENT TEMPERATURES AND THICKNESSES OF CAN ICE (HOURS).

| Thickness. | 1 in. | 2 in. | 3 in. | 4 in. | 5 in. | 6 in. | 7 in. | 8 in. | 9 in. | 10 in. | 11 in. | 12 in. |
|---|---|---|---|---|---|---|---|---|---|---|---|---|
| Temp. 10° F. | 0·32 | 1·28 | 2·86 | 5·10 | 8·00 | 11·5 | 15·6 | 20·4 | 25·8 | 31·8 | 38·5 | 45·8 |
| ,, 12 ,, | 0·35 | 1·40 | 3·15 | 5·60 | 8·75 | 12·6 | 17·3 | 22·4 | 28·4 | 35·0 | 42·3 | 50·4 |
| ,, 14 ,, | 0·39 | 1·56 | 3·50 | 6·22 | 9·70 | 14·0 | 19·0 | 25·0 | 31·5 | 39·0 | 47·0 | 56·0 |
| ,, 16 ,, | 0·44 | 1·75 | 3·94 | 7·09 | 11·0 | 15·8 | 21·5 | 28·0 | 35·5 | 43·7 | 53·0 | 63·0 |
| ,, 18 ,, | 0·50 | 2·00 | 4·50 | 8·00 | 12·5 | 18·0 | 24·5 | 32·0 | 40·5 | 50·0 | 60·5 | 72·0 |
| ,, 20 ,, | 0·58 | 2·32 | 5·25 | 9·30 | 14·6 | 21·0 | 28·5 | 37·3 | 47·2 | 58·3 | 70·5 | 84·0 |
| ,, 22 ,, | 0·70 | 2·80 | 6·30 | 11·2 | 17·5 | 25·2 | 34·3 | 44·8 | 56·7 | 70·0 | 84·7 | 100·0 |
| ,, 24 ,, | 0·88 | 3·50 | 7·86 | 14·0 | 21·0 | 31 5 | 42·8 | 56·0 | 71·0 | 87·5 | 106·0 | 126·0 |

In the early days of artificial ice-making it was sought to produce clear can ice by freezing slowly at a high temperature—25° or thereabouts. The ice produced was clear, but the time taken was so great that the idea was abandoned in favour of the cell system and agitated ice. If, on the other hand, the rate of freezing is too quick, the ice is not close-grained, but consists of comparatively large crystals loosely combined. This affects adversely the keeping quality of the ice.

**Capacity of Tanks.**—The capacity of an ice-tank is usually stated in tons per 24 hours, but as, with the exception of small cans, can ice will not freeze in 24 hours, but sometimes takes as long as 48 hours, it is necessary to have a greater number of cans in the tank than are required to give the nominal output per 24 hours. Obviously, if the ice in the cans in a tank could be frozen solid in 20 hours from their being put in, and each can contained 112 lbs. of ice, only 400 cans would be wanted to give a nominal output of 20 tons of ice per 24 hours; but a 112-lb. can requires longer than 20 hours in which to freeze, so that a greater number of cans are required. There is another very important reason why more cans than those actually required to give the nominal output should be put in. If in the above case the 400 cans were all filled with water at the same time, the machine would be doing its maximum work at the beginning; as the ice froze, this would gradually fall off, until, when all the cans were frozen, it would be doing no work at all. The essence of economical work is to have regular work, and the load on the machine should be kept as regular as possible; to ensure this some cans should be freezing while others are being harvested. In the United States, where distilled water is almost universally used, the cans are usually filled and dumped separately and one at a time; this tends to ensure regular work, but gives more labour.

With 112-lb. cans it is usual to put in from $1\frac{1}{2}$ to $1\frac{3}{4}$ the nominal number of cans. This ratio varies with the rate of freezing and with the size of the cans. For 200-lb. cans (11-inch ice) it is quite usual to put in as many as $2\frac{1}{2}$ times the number of cans that the nominal tonnage of the plant would require.

**Piping.**—The surface of the piping in the tank varies with the size of the plant. The pipes run along between the cans, and are usually of 1-inch or $1\frac{1}{4}$-inch pipe. Each coil, which has from 850 to 950 feet run of pipe, is fed by an expansion cock connected to the liquid header at the top of one end of the tank. The suction end of this coil is connected to the suction header either at the top of the same end or at the other end, as is most convenient. Both top and bottom feed are used by different makers. In some cases a combination of top and bottom feed known as top feed and bottom expansion is used. In this case the coils are fed with liquid through expansion valves connected to a liquid header running along the top of the back end of the tank. The liquid runs down through the coils as in ordinary top feed until it gets to the bottom, when it passes into an equalizing manifold, and expands up again through another coil which is connected to the suction manifold running parallel to and a little behind the liquid manifold. It will thus be seen that each alternate coil is a top and a bottom feed. The system combines the advantages of each, and has none of the dangers.

For 112-lb. cans of 8-inch ice 33 inches deep there are usually 6 coils of pipes. For 200-lb. cans, 11 by 22 by 36, there are 8 coils of pipes.

**Circulation of Brine.**—Brine-tanks should be made long and narrow. A 20-ton tank, for instance, containing 300 2-cwt. cans would be 16 feet wide, 40 feet long, and 4 feet 6 inches deep.

It is of paramount importance that the brine in a tank should have an even temperature throughout its length, breadth, and depth. It must be remembered that at low temperatures the specific gravity of brine only varies very slightly, and as on this account natural circulation, even if it were good enough, cannot in any way be depended upon, forced circulation must be resorted to.

Brine-tanks should be made with a partition down the middle, and the best circulating device is a propeller running astern—*i.e.*, sucking the brine and running at a speed of from 100 to 130 revolutions per minute.

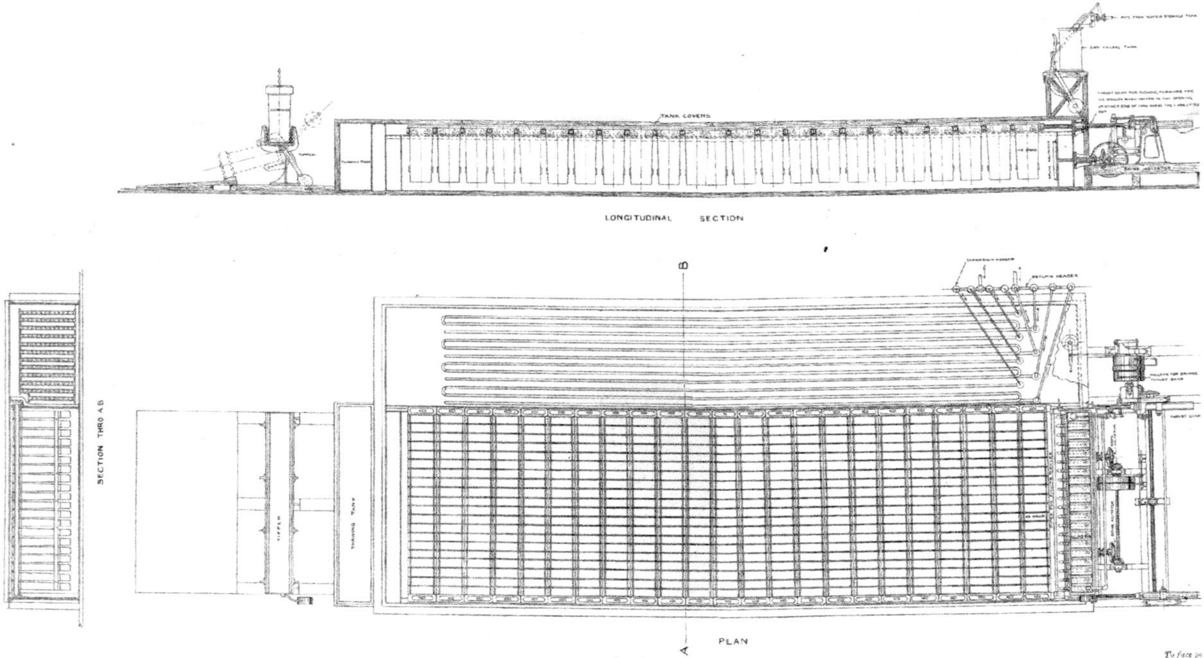


Fig. 85.

To face p

The tank should be made with baffles at each end, so as to equalize the circulation as much as possible. Sometimes a centrifugal pump is used, sucking the brine from the bottom of the tank and delivering it over the coils in the tank through perforated pipes which run the whole length of the tank. Pumps are always used when an external cooler is employed, but for coils inside the tank a propeller gives quite satisfactory results. A law previously quoted in connection with condensers (Chapter VI.), stated that the transmission of heat from a hot body to a cold body depends on the rate at which the substance to be cooled is passing over the cooling surface. This law holds good for ice-tanks, and the better the circulation, the more even will be the temperature and the quicker the rate of freezing. With even temperatures also, higher suction pressures can be carried than when the circulation is bad.

**Operation of a Can Plant.**—The operation of a can plant consists of five distinct steps. These are:—

(*a*) Filling the cans with water.
(*b*) Lowering the cans into the brine.
(*c*) Lifting the cans out of the brine when frozen.
(*d*) Thawing off or loosening the ice in the cans.
(*e*) Dumping the ice, or emptying the cans.

In some plants the cans are first lowered into the brine and then filled with water.

In order to keep the brine at an even temperature and also an even load on the refrigerating machinery, the process of filling the cans with water and harvesting the ice should, as far as possible, be continuous.

Fig. 85[1] illustrates a can ice tank in plan and section. The tank belongs to the second general class given on

[1] Messrs L. Sterne & Company, Ltd., Glasgow.

page 275, because the cooling coils are placed in a separate part of the tank to the cans. It is claimed that by this arrangement a better circulation of brine at a more even temperature can be obtained than by any other method, and that the ice-cans can be put more closely together, thus reducing the size of the can-fillers, ice-dump, and thawing-off tank.

The tank illustrated is capable of turning out 25 tons of ice per 24 hours. It contains 460 cans, each 42 inches by 17½ inches by 6 inches, and each block of ice weighs 140 lbs. Each row of cans is carried on a frame extending right across the tank. The ends of this frame rest on rollers running on an angle-iron track riveted to the side of the tank about 12 inches from the top as shown. These frames are lifted by the crane so that a whole row of cans, twenty in number, is moved, filled, and emptied at one time. To a certain extent this is a difference between British and American practice. It is common in the States to rest the cans on the bottom of the tank and to provide each can with a separate wooden cover; in this case, as before noted, the cans are filled and emptied separately. In some of the more recent plants the British practice has been adopted, a single cover, extending right across the tank, being used for each row of cans. Where distilled water is used almost universally as it is in the States, and each can is filled separately by a special filler and immediately covered up, there is a good deal to be said for the single covers, as they are undoubtedly more air-tight as regards each can. It is also urged in favour of the separate cover that, being light, it can be readily removed by hand, whereas with a large cover which is shifted by power there is danger of dirt from the cranes getting into the cans. This latter argument, however, is not one that can be put forward with any force, as dirt is apt to lodge in the cracks between the single covers, and dirt from the crane, if any, will be just as likely to drop into one can as into a dozen.

The circulation of the brine in the tank is caused by two propellers placed at the back end of the tank. These are kept running astern, and the brine is drawn by them from the freezing tank and forced into the cooling tank, where it comes in contact with the ammonia expansion coils. The communication between the two tanks is always open, and means for regulating the circulation is not found necessary. As the brine is drawn from the freezing tank at one end, its level in this tank falls, and the cold brine from the cooling tank flows into the freezing tank through an opening in the division between the two tanks placed at the end of the tank farthest from the propellers, thus causing a constant circulation.

It will be noticed that top feed is used in the expansion coils. As the ice becomes frozen at the front end of the tank, the end row is lifted out by the crane, lowered into the thawing tank, and then hoisted on to the dump; the frame carrying the moulds remains hooked on to the crane the whole time, and, on the dump being tipped, the twenty blocks of ice are discharged from the cans. The row of cans is then raised from the dump and carried to the back end of the tank by the traveller, which stops automatically when the proper position for filling the cans from the filling tank has been reached. At this point, two rams, which have a stroke equal to the width of a row of cans, and which are mounted at the back of the tank, are put into motion by the attendant, who starts the traversing gear by the handle which is shown carried on a pillar fixed to the cover of the brine-cooling tank.

These rams, acting on the back row of cans, push the whole of the cans in the tank forward until the space in the front end is filled up and there is a space at the back of the tank ready to receive the cans just removed as soon as they have been filled with water. When the cans have been pushed forward the requisite distance, the traversing gear is automatically reversed by a striking gear on the ram, and the rams are returned ready for the next row. It will be seen that this process is continuous. As soon as a frame is lifted out, which occurs at regular intervals of time, the rest are moved up, and the empty can is filled and replaced in the tank at the back. Each can in a frame is equally frozen, and each successive row from the back contains more and more ice.

To avoid confusing the figure, the agitating gear has not been shown.

Instead of lifting a separate frame for each row of cans, some makers use a single frame made of steel joists, from which are suspended chains coinciding in position with the eyes on each end of the cans. When the ice is to be harvested, the cans are hooked on to this frame and a whole row lifted at one operation. With this arangement, however, the chains are in the way when the ice is being dumped. A modification of, and an improvement on, this is to have the cans set in a light frame provided with, say, six ring bolts. When the ice is drawn, the chains are hooked into these eyes; the chains are not in the way, and the cans are not interfered with.

**Thawing-off Tank.**—When the cans have been drawn from the tank, they are at a temperature of, say, 15° F., and the ice in them is adhering closely to the metal. In order to enable the ice to be withdrawn from the cans it must be loosened. This is done by immersing the cans in a tank, the water in which is kept at a temperature of about 70° F. by a steam-coil. As the metal gets hot it thaws the ice next it, and the block of ice in the can is loosened.

The thawing-off tank usually extends right across the entire front of the ice-tank as shown in fig 85, and is just large enough to take a whole row of cans at once.

The tank should have a broad white band painted on the inside to show the level of the water (when the tank is idle) that will just suffice to nicely surround the cans when they are immersed in it. If this is not done the tank may get filled too full, and the displacement of the cans will cause it to slop over and wet the insulation of the freezing tank. In any case the thawing-off tank should be provided with a large overflow, and there should, of course, be a waste-cock and a supply-pipe to make up the waste, so that the tank can be emptied and filled with fresh water occasionally.

These tanks must be perfectly free internally, all staying being external. As they are usually incased in wood, this is an easy matter to arrange.

**The Spray Thawer.**—A system of thawing off commonly used in the United States is known as the hot spray. With this the thawing-off tank is dispensed with and the cans are laid on the tipping table as they come direct from the ice-tank. While in this position water at a temperature of 130 to 160° F. is sprayed over them. The thawing-off action, as well as the sudden expansion of the metal due to the heat, loosens the ice and it shoots out on to the handling platform. This method has the merit of being rather quicker than the thawing-off tank, but the difference is only one of minutes and is, comparatively speaking, not important. The wear and tear on the cans is greater, and if they are riveted they are liable to become leaky.

**Ice-Dumps.**—The ice-dump is one of the most simple, though one of the most necessary, portions of the plant. A handy, well-balanced dump will save an enormous amount of labour, besides increasing the life of the cans by preventing their being knocked about. Where the cans are handled singly, nothing very special in the way of a dump is wanted. A good single-can dump is manufactured by the Triumph Company,[1] and consists of a metal frame worked on a trunnion. The spray-pipe is fixed round the top edge and to the bottom of the dump, and is connected to the supply by a three-way valve. The action of tilting the dump turns on the water, and when the ice has been thawed out the dump is righted and the spray turned off.

A balanced dump such as is used for dealing with a large number of cans at one time is shown in fig. 85. The cans are shown in position, being held there by the crane overhead, and the position the dump assumes when tipped is shown in dotted lines. The bottom and sides of the dump are lined with wood to prevent injury to the cans.

The action of a dump of this class is admirably shown in fig. 86.[2] The dump has just been tipped and the opaque ice is just issuing from the cans, which are held in position by the traveller overhead. The photograph is an admirable one and shows the action very distinctly.

Though balanced, it is very necessary to prevent the dumps swinging over with a rush, and for this reason the cans are placed in the dump by the crane, and the rope is gradually slacked out as the dump swings over. The dump in this way is gradually lowered into position, and the ice dumped. When relieved of the ice, the dumps will either right themselves automatically, or they can be

[1] Messrs the Triumph Ice Machine Company, Cincinnati, O., U.S.A.
[2] Messrs the Linde British Refrigerating Company, Ltd., London.

FIG. 86.

*To face page* 286.

easily righted by hand. Owing to the weight they have to carry, the trunions and frames want to be substantially built and firmly secured to their foundations.

**Can-Fillers.**—The type of filler used depends on whether ordinary water or distilled water is being used in the cans. If distilled water, fillers of the type shown in fig. 79, page 248, have to be used, so as to prevent the water becoming aerated. If ordinary water is being used, aeration does not matter so much, and the filler then employed usually consists of a long narrow tank about the same size as the thawing-off tank, but divided into as many compartments as there are cans in a frame. These compartments hold exactly a can full. They are filled by a large pipe direct from the forecooler tank, or, if this is absent, from the water main. From each compartment a pipe usually connected to the tank by a flexible hose to allow freedom of movement leads to its can. The outlet from each compartment is controlled by an ordinary draw valve connected to a common spindle and handle, so that all the compartments in the tank can be emptied, and all the cans filled, simultaneously. In some designs these filling tanks have the outlet to the supply-pipe to the cans kept 2 inches above the bottom of the tank, the idea being to prevent any settlement that may take place being drained into the cans, a purge-cock being provided for cleaning out this 2 inches of water at intervals. There is, however, no particular merit in this arrangement. If the water being used is pure there will be no settlement, and if there is settlement the sediment will get stirred up every time the tank is filled. It is not advisable to have a dead water space in which it is possible for stale water to collect. Purge-cocks only add to the expense; the outlet pipes can be quite well put in flush with the bottom of the tank so as to draw all the water out, or fixed with a back nut in the ordinary way.

In order to dispense with the draw valves—in fact any valves at all—an improved form of filling tank has been introduced. This is shown in fig. 84 and consists of a number of pipes, each connected by flexible pipe to its own compartment. These pipes are fixed, at centres to suit their cans, on a light angle-iron frame, which is attached to a rocking shaft, and is capable of being revolved through a quarter of a revolution by it.

When the tank is full the pipes and frame are raised vertically against the side of the tank as shown, the outlets being some 5 or 6 inches above the water-level. On the shaft being rocked by the crane man, the pipes quickly assume the horizontal position, each pipe over the centre of its can, and empty the contents of the compartment into the can. It is important that the rocking motion should be done smartly or the water will spill, and also that each compartment should hold no more and no less than the exact quantity of water required to fill each can to its proper level. Fig. 87 [1] shows a can-filler at work filling its twenty-eight cans simultaneously. The tank behind is divided into twenty-eight compartments, one to each can; and each compartment when full holds no more and no less than the exact amount of water required to fill the can. The cans are suspended above their place in the ice-tank; when full, they will be lowered into the brine and the wooden covers replaced.

**Agitation of the Water in a Can Ice Plant.**—Various apparatus have from time to time been used for agitating the water in a can ice plant with the object of depriving it of air and enabling it to freeze clear. In 1901 Mr

[1] Messrs the Linde British Refrigeration Company, Ltd., London.

FIG. 87.

*To face page* 288.

Skinkle read a paper before the Western Ice Manufacturers' Association,[1] in which he described two new systems of agitation, compressed air being used in both cases.

In a plant belonging to the Simon Ice Company of Louisville, Ky., the discharge pipe from the air-blower is connected to a manifold leading along the front of the tank. From this manifold pipes are led back between each alternate row of cans. These pipes are connected by a cock with a small rubber tube, which hangs down the centre of the can to within a few inches of the bottom. When the can is full of water and freezing is commenced, the air is turned on, and, issuing from the bottom of the rubber tube, produces violent agitation of the water. The air bubbles as they ascend wash the freezing surface and dislodge the air and other impurities. When the ice is almost closed the rubber tube is taken out, and the water remaining unfrozen, and which is said to contain all the impurities, is removed, and the empty space filled up with distilled water. In this plant a circulation of about $\frac{1}{10}$ of a cubic foot of air per minute was used for each 300-lb. can, and the results are said to have been exceedingly satisfactory.

The other system was invented by Mr Ullrick, of the National Refrigerator Company. In this the compressed air, after being washed and dried, compressed and expanded, reaches the cans at a temperature of about 8° F. Each can in the tank has a galvanized pipe 1 inch in diameter passing down the outside of the can and terminating in the centre of the bottom of the can, which is concave and fitted with a special fitting having a rubber ball valve, so as to prevent water from the can entering the

[1] *Ice and Refrigeration*, vol. xx., No. 4, p. 159.

air-pipe. Inside the 1-inch pipe there is a small air-pipe connected to the ball valve at the bottom and to the air supply at the top. This small pipe conveys the air, and the space between it and the inch-pipe is a dead air space provided for insulating purposes.

The air from the compressor enters the can at the bottom through a hole about $\frac{1}{16}$ of an inch in diameter, and the circulation is kept up until the hole is frozen over. The impurities collect at the top of the ice block, and are removed before the ice is sold.

In another method[1] of freeing the water from air a small air-pump is fitted with a differential action, so that the delivery stroke is made at twice the speed of the suction. The delivery pipe is connected by pipes to glass vessels shaped like a bottle and suspended over the centre of each ice-can. These glass vessels have wooden nozzles which dip into the water in the cans. On the suction stroke of the air-pump sufficient water to fill it is sucked up into the glass bottle. On the delivery stroke—at twice the speed—this water is violently expelled, and, owing to its velocity, passes down to the bottom of the can. The agitation thus produced thoroughly washes the surface of the ice as it is forming. In practice the cross tubes carrying these bottles are supported on wire ropes running along each side of the tank and fixed to a rocking shaft at the back end of the tank. This shaft is provided with a counterbalance, and by rocking the lever the wood nozzles can be lifted clear of the cans when they are thrust forward by the rams.

[1] Messrs the Linde British Refrigeration Company, Ltd., London.

## THE PLATE SYSTEM.

Plate ice can be produced in two ways:—

(a) By freezing from one side only, or

(b) By freezing from both sides, the ice meeting in the middle.

In order to produce clear hard ice in a plate system, it is not necessary to distil the water. There is little danger of air being frozen in as long as it is able to escape and has sufficient time in which to do so. Plate ice generally is harder than can ice and lasts longer.

The great drawback to plate ice is the length of time it takes to freeze, 11-inch ice, which is the usual thickness, requiring as much as ten days. This necessarily means a very large plant for a given output, and though the plant is simplicity itself, the first cost is higher than that of a distilled water can plant, and the space occupied is also greater—an important consideration.

The tank can be of any length, depending on the output. It is usually about 14 feet wide and from 6 to 7, or even 10, feet deep. The freezing plates or hollow walls do not extend right across the tank, but room is left for expansion. The main tank is divided into a number of distinct compartments or small tanks by wooden partitions, and all the plates in any one section freeze at the same time.

When the tank is filled with water and the brine[1] circulation started, ice begins to form on the outside of the walls, becoming thicker and thicker as long as the brine supply is kept up.

The plates were originally made of cast-iron about $\frac{3}{8}$ of

[1] In some plants direct expansion is employed in the freezing plates instead of brine. Harvesting is done by turning the warm ammonia gas through the coils.

an inch thick, and divided into about ten passages. Great difficulty was experienced with this design as the cast-iron plates cracked, and the plates are now made either of wrought-iron galvanized plate $\frac{3}{16}$ of an inch thick, kept apart by distance pieces and riveted together in the form of a hollow wall as shown in fig. 91, or of coils of pipe covered on either side with a thin sheet of iron riveted through. This last method is sometimes preferred, as the brine circulates continuously through the coil, and there are no dead corners as are sometimes found in the plain wrought-iron walls. Care must be taken that all rivets are countersunk or the ice will freeze round them, and it will be exceedingly difficult, if not impossible, to harvest it. The water between the two plates must be continuously agitated by long thin wooden paddles or by compressed air. As soon as the ice has reached the required thickness, the remaining water, which contains all the impurities, is run off; the supply of cold brine is turned off, and warm brine from a special tank, where it is heated by a steam-coil, is turned on and allowed to flow through the coils, heating them and thawing the ice. As the warm brine enters the coils, it naturally drives the cold brine before it. This continues to flow to the storage tank until it has all been displaced, when the cock is shut and the warm brine is returned to its own tank. When freezing is resumed, the same process, in the opposite direction, is followed. The warm brine is cut off and the cold brine turned on.

Unless the agitation is very efficient, there is a liability for the ice to creep and freeze across the bottom of the space between the two plates. If this occurs and the ice-plates unite, it is impossible to withdraw the ice without cutting it with a steam-pipe. In order to avoid this difficulty,

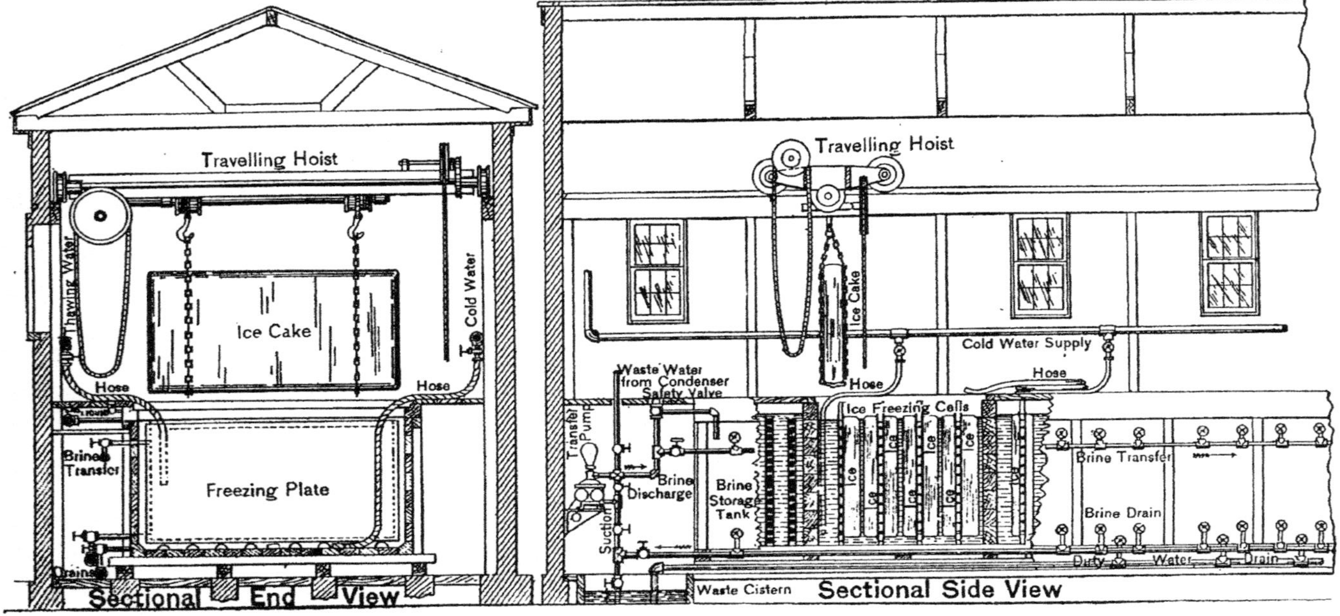


FIG. 88.

*To face page 293.*

a hollow square iron trough is sometimes placed in the water space between the two blocks. If the plates are 27 inches apart and the ice is to be 11 inches thick, the trough will be 5 inches wide. The trough is connected to the warm brine supply. When freezing is in process it is full of cold brine, but there is no circulation. When freezing is completed warm brine is turned on, which displaces the cold brine, thaws off the ice at the bottom and side, and allows the cake to be drawn out with a square bottom and square edges.

Fig. 88 [1] illustrates a plate tank; the general idea will be seen at a glance, as the figure is self-explanatory.

**The Block System**—In the plate system the ice never meets or freezes together unless by accident. Attempts however, have been made (system (*b*)) to freeze from both sides, letting the two plates meet and unite in the middle to form a solid block. This is known as the block system, but has not come into general use. The plates are fixed about 12 inches apart, and 6 inches of ice is frozen on each side. As the ice freezes together there is no room for expansion, and difficulties are met with from this cause. The most serious disadvantage, however, consists in the core which forms at the point of junction of the two plates. The ice will not freeze together and make a clean junction, as the air in the water will all collect in the centre and prevent it. The core shows up very badly when the ice is melting, and it has a material bearing on the rapidity with which liquefaction takes place. Block ice is also difficult to withdraw from the tanks, as there is no room to assist thawing operations by prising the ice out of place.

[1] Messrs the Frick Company, Waynesboro, Pa., U.S.A.

## THE CELL SYSTEM.

While the can system produces ice in tapered blocks from 6 to 8 inches thick, and the plate system in large plates about 11 inches thick and weighing several tons, the cell system produces it in nearly square blocks, and as a cube practically[1] presents the smallest external surface for the maximum capacity, ice in this shape melts less readily than in any other.

Theoretically the cell system will produce blocks of perfectly pure ice. Actually the ice is seldom so clear as the natural product, and unless the greatest care is exercised in its manufacture, it will freeze with a core which detracts from its appearance.

The cell tank is usually made of hard yellow or Baltic white pine, the planks being bedded in hot asphalt before being bolted together by long bolts passing through them. The ends and sides are built in the same way, the sides being housed into the bottom and ends, and the whole made perfectly water-tight.[2] The tanks can be made in any size, but usually run about 10 feet square and 5 feet deep. The tank is provided with a false bottom which forms the bottom of the ice cells.

A 10-feet by 10-feet by 5-feet tank will contain about twenty cells, the sides and ends of which are formed of hollow iron walls supported on the false bottom. Through these walls the brine circulates, freezing the water in the cells on each side. The false bottom is pierced with a number of rectangular holes, one coming in the centre of the bottom of each cell. In addition to the false bottom

[1] Actually, of course, this condition is fulfilled by a sphere.

[2] In some plants the planks are jointed together with thick waterproof paint, the vertical joints being made with rushes.

there is a false side, and between this and the real side there is a space in which a paddle or plank, worked off a rocking lever by the engine or by an electric motor, rises and falls. When the whole tank has been filled with water, the brine is turned on and freezing commences. At this juncture the paddle is started, and, by rising and falling in the closed space at the side, causes the water to gently flow backwards and forwards through the holes in the bottom of each cell. By this agitation the water keeps washing the freezing surfaces free from air bubbles and impurities, and allows the water to freeze clear. The agitation is continued until the water in the cell is all but frozen solid; it is then stopped, and the whole block is allowed to solidify. The water between the false bottom and the bottom of the tank is not frozen, and is drained off before the ice is harvested.

The thawing-off process, which is a slow one, taking from two to three hours, is very much like that adopted in the plate plant, and consists of turning off the cold brine and circulating warm brine until the ice is loosened in each cell. When loosened, the blocks are lifted out of the cells by the crane, which engages with a piece of rope or an iron ring which has been frozen into the ice.

A cell plant is more costly to put down than a can plant of equal capacity.

Figs. 89, 90, 91, and 92[1] show, in section and plan, details of a cell tank containing twenty-four cells. In a complete plant these are placed in rows one after the other until the required capacity of the plant has been reached. All the cells in a tank are frozen at the same time, but, as in the plate plant, freezing in the separate tanks can be at different stages, so that while the ice in one tank is being

[1] Messrs L. Sterne & Company, Ltd., Glasgow.

harvested, that in the others will be freezing. This keeps a regular load on the machinery, and enables ice to be produced at regular intervals. The figures show very clearly the construction of the tanks. The circulation of the brine through the cells is shown by arrows. On reference to the plan, it will be seen that the cold brine

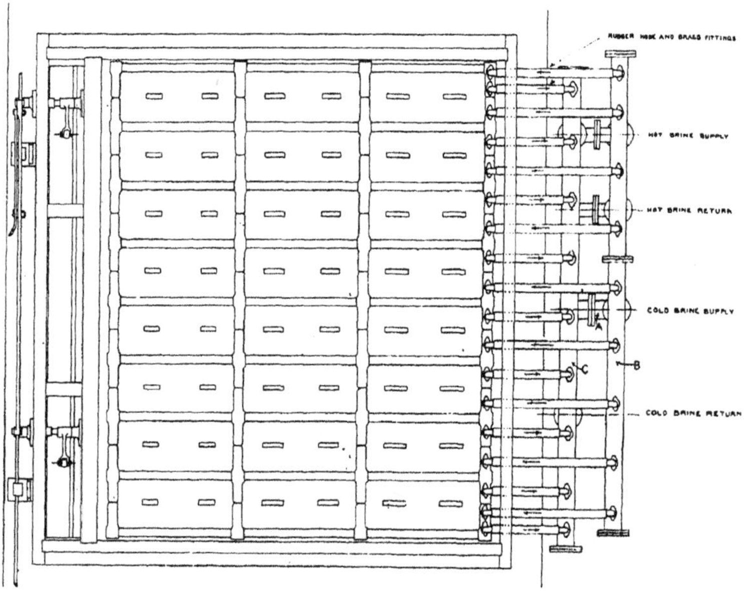


PLAN

FIG. 89.

from the cooler passes from the main supply-pipe *A* to the distributing header *B* placed at the side of the tank. From this it is taken through rubber pipes to the cast-iron box at the end of each cell, whence it pursues the zigzag course indicated by the arrows, shown in fig. 90, being baffled backwards and forwards, and finally returns

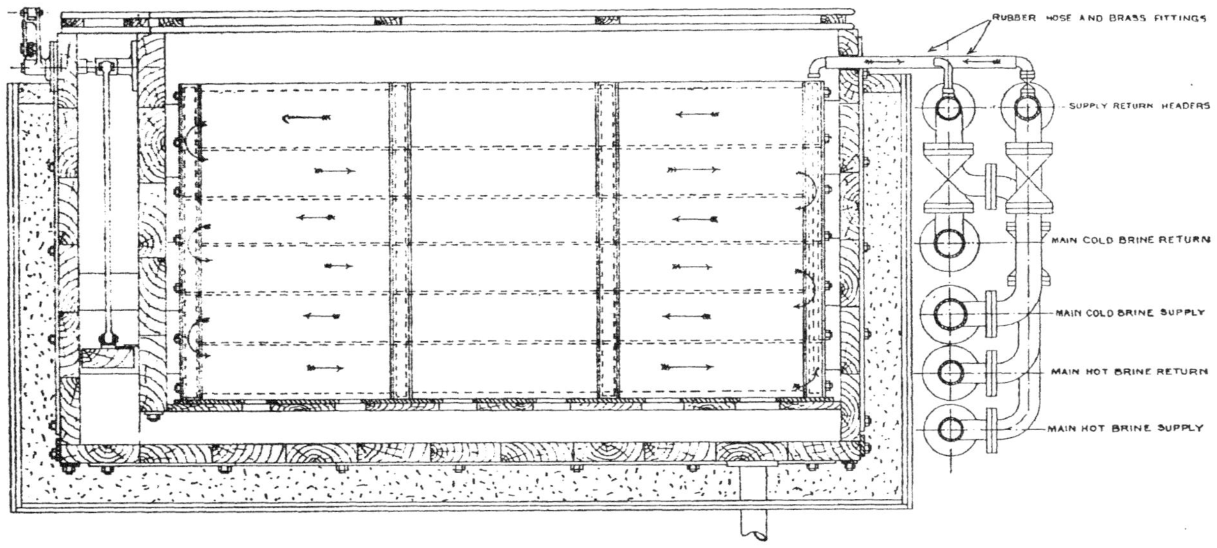


TRANSVERSE SECTION

FIG. 90.

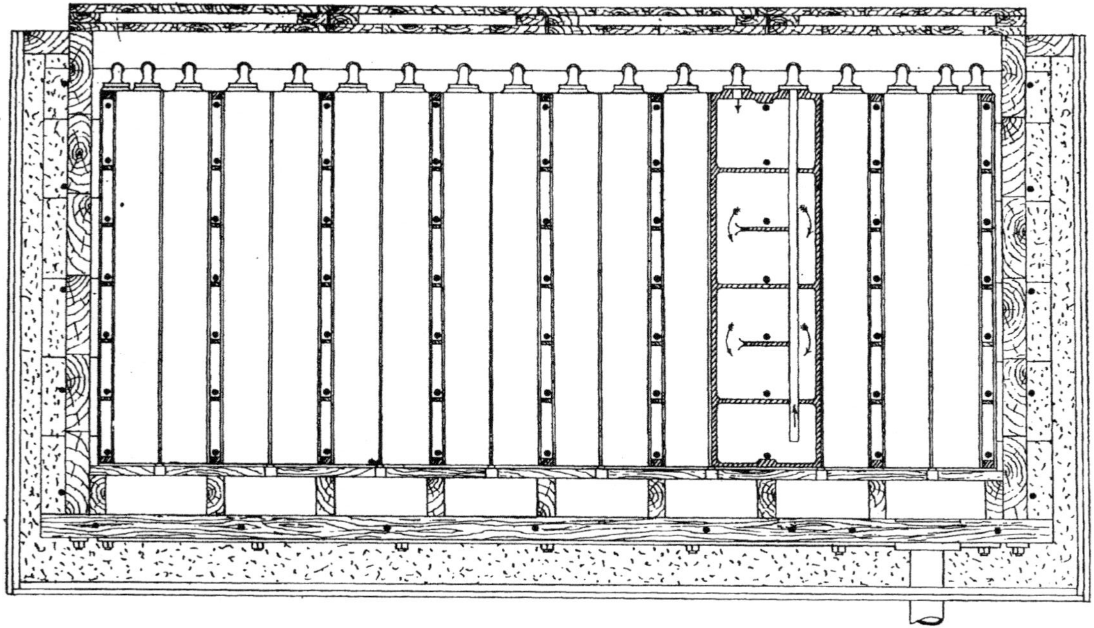

LONGITUDINAL SECTION

Fig. 91.

to the cold brine return header *C* through the $\frac{7}{8}$-inch diameter pipe *D*, fig. 92.

When the ice is frozen and is required to be thawed off,

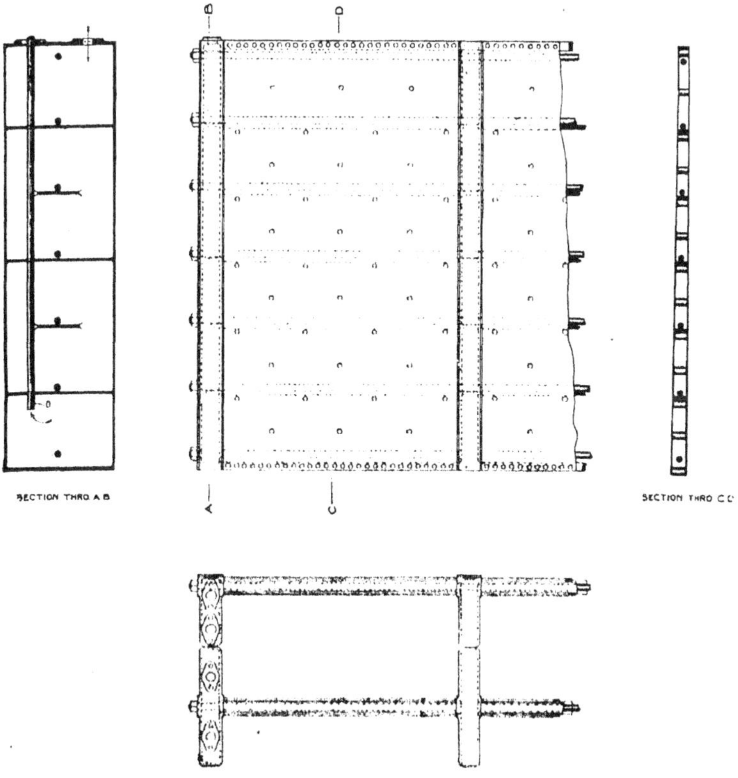


FIG. 92.

the main valve in the cold brine supply-pipe is shut, and that in the hot brine supply opened. The hot brine now pursues an opposite course to the cold brine. It enters the cell by the pipe *D*, and flows to and fro until it returns through the top of the cell to the header *B*. At this point the cold brine return is shut off and the warm brine return opened. No mixture of the cold and hot brine takes place, each returning to its proper tank.

Each wrought-iron brine wall or cell is made of $\frac{3}{16}$ of an inch plates riveted to bars top and bottom as shown in fig. 92. There are also a number of stay rivets through the cells to support them. Each cell is independent of the other so as to allow for expansion, and two holes are allowed in the bottom of each cell instead of one, to ensure a better agitation of the water. The agitating paddle is shown at the side of the tank in the transverse section.

Freezing with the cell system is a slow process, and a block weighing, say, 6 cwt. will require eighty hours in which to solidify.

**Storage and Sale of Manufactured Ice.**—Ice-making is a seasonal business, and a point which is frequently discussed is whether it is most economical to put down a plant sufficiently powerful to make all the ice required in the season and to shut down entirely for the idle months, or whether it is best to put down a small plant with an ice store and manufacture the year round, storing the ice made in the winter and selling it in the summer when the demand is high.

Briefly, the arguments for the former method are that ice will melt and deteriorate in store, and that the cost of storing the ice, and the double handling which this necessitates, is as great, if not greater, than the interest on the larger plant. Also, that when the plant is shut down the men can be discharged, and as no wages are being paid or coal burned the outgoings are small.

For the latter system it is argued that in the first

instance a much smaller and cheaper plant is required, and that when once the hands have got used to a plant and are able to handle it to the best advantage, it is unwise to discharge them and get in new men every season; that the machinery is apt to deteriorate if it is left idle for long; that the cost of making ice continuously and storing it in properly constructed refrigerated rooms is not so great as the interest on the capital invested in the larger plant; that to always have a store of ice to draw on when the demand is heavy and the prices are ruling high is a great convenience and prevents the tanks being robbed; and that it is cheaper to make ice in the winter when all the natural conditions are in favour of a low temperature.

When opaque ice is being made and sold for special purposes, such as the supply of the fishing-fleets, the question is different; but in the majority of cases, and for general trade, there can be no question as to the superiority of the latter system. This is especially true when an ice factory is being worked in conjunction with a cold store, and where the same machinery and same staff is being used for both.

Where the cold stores rely for their business on produce sent them by merchants, the storage room is rarely fully occupied in the winter months, and under these circumstances a large quantity of the ice made can, if the floors are properly constructed, be stored in the vacant space. In the spring the demand for ice begins simultaneously with the demand for cold storage space, and as the produce comes in the ice goes out.

**Position of Ice Stores.**—It is always desirable, however, not to rely upon the produce stores for ice storage, but to have, in addition, special ice stores which are regularly used for the purpose, and which take the bulk of the output, the produce stores being used for the overplus. In most towns considerable excavation has to be made for foundations, and in this case it is best to make the ice stores in the basement, the tank being on the ground floor, so that the ice can be passed direct from the dumping platform to the store. This position also has the advantage that the only opening into the ice store is at the top, and that there is no tendency for warm air to enter it. Circumstances, however, are not always propitious, and frequently the store has to be on the ground floor.

It is advisable, if possible, to gravitate the ice into the store, and in this case the ice tank can well be placed on the first floor, the store being either directly under it or adjoining it on a lower level.

Ice stores should be capable of being ventilated, and should be constructed with an ante-room capable of holding two or three days' supply. The use of this room will avoid the necessity of opening the main store at frequent intervals, and so will save refrigeration. Both the main store and the ante-room should be well insulated and piped with direct expansion or brine pipes, so that the temperature will be always about 22° F.[1]

As the ice is naturally always at a low temperature, the amount of piping required can be about one-third less than that required for the produce stores. Where the ice store is situated directly below the ice-tank, the bottom of the latter is sometimes left bare and is used to cool the store. The storage capacity required is from 50 to 55

[1] If the room is kept at a higher temperature the ice will become soft and will readily melt when taken out of store. In the United States the stores are sometimes kept at 16° F., and ice for storage purposes is frozen, with the brine at a specially low temperature.

cubic feet per ton of 2240 pounds. This allows for handling room, but necessitates the ice being closely packed.

**Construction of Ice Stores.**—Generally speaking, the construction of an ice store follows the same lines as that of a produce store, the walls being of brick. Efficient insulation is the primary consideration, as by its means the wastage of the ice will be some 20 per cent. less than that of an uninsulated store. The feature of an ice store requiring special design is the floor. This should be insulated, as heat entering from the ground tends to make the bottom layer of ice rotten and porous.

It is best constructed as follows:—Over the whole of the floor spread concrete, the thickness depending upon the nature of the ground; this should be laid with an easy fall to a gutter, which is frequently placed in the centre. Upon the concrete lay 9-inch by 2½-inch joists with sound boarding as shown in fig. 75; fill in the space above these sound boards with insulating material, and cover the joists first with 1-inch rough boarding, then with one or two layers of strong waterproof paper, the top layer being laid one half lap over the other. On this paper lay 2-inch by 4-inch or 2-inch by 6-inch boarding, calk the seams between these planks with oakum, and run in with pitch as in a ship's deck. The walls may be insulated as for a produce store, but the outer board should be at least 1 inch thick. The paper in the floor should be carefully flashed up at least 12 inches and joined to that in the wall, and the junction of the wall and floor should be protected by a 9-inch by 2½-inch skirting-board with the arris taken off. This should be housed to the floor, and the seam between it and the floor should be calked and pitched. This makes a perfectly watertight floor, with a fall similar to that of the concrete.

The gutter can be of 7-lb. lead, the joists in the centre being shaped to receive it; the lead must be flashed up under the 6-inch by 2-inch flooring. The gutter should be laid with a fall towards the centre or one end of the room, and must drain into properly trapped gulleys. The gutter area need not be great, as, if all goes well, the drainage will be *nil*, and at the worst will be small. On the floor, and run in the direction of the fall, 3-inch by 2-inch timber should be laid. On these the blocks of ice rest. The 3-inch space between the bottom of the ice-block and the floor allows a circulation of air under the blocks, and also allows any drainage that may take place to easily run off. Fig. 93[1] is an illustration of the ice storage room at the Mound City Ice and Cold Storage Company's premises, St Louis, Mo., U.S.A., and shows generally the design that should be followed.

Some designers make the ice store floor of 2 feet of coke breeze or cinders and do not insulate it. This construction, however, is not general or good.

The 3-inch by 2-inch timber should be so placed that each block of ice will rest on three strips, one at each end and one in the centre. These timbers should meet over the gutter, so as to prevent the lead being injured. It is recommended that the first tier of ice should be set on edge (not on end) with a 2-inch by ½-inch lath strip between each block and its neighbour, both vertically and horizontally. With can ice the blocks should be laid with the big and little ends alternating in each layer. If the ice is carefully packed in this way with plenty of room for ventilation, and the store is refrigerated, there will be no tendency for the blocks to freeze together. In the United States rice straw is

[1] Reproduced from *Ice and Refrigeration*, Chicago.

Fig. 93.

*To face page 304.*

packed between the blocks, and takes the place of the wooden laths.

**Handling of Ice in Store.**—If the ice-blocks are of small size, all necessary moving and packing can be done by hand. With 200-lb. or 300-lb. blocks, however, if there is more than one tier this is impossible, and the blocks have to be moved and packed mechanically. In large stores an electric traveller can be used, the blocks being picked up as they enter the store and placed in position by it. In America portable pneumatic hoists are used. These are fitted with a flexible air supply pipe, and are moved about and taken from room to room as required. They may also be made to run on a track secured to the ceiling, two or three tracks being fixed in each room, and the hoist moved from one to the other according to the particular portion of the room requiring to be dealt with.

**Transport of Ice.**—It is a difficult matter for manufacturers to urge the purity of their product with any prospect of being believed, when the consumer sees the ice being carted about and delivered at an hotel, his door, or the fishmonger's in a dirty waggon by dirtier men. In the method of delivery now commonly in use, a dirty pad is placed on the ground below the tail-board of the waggon; on this the ice is dropped, and is then dragged across the road through the gutter, over the pavement, and into the shop, where, after being sluiced down with water, it is offered for sale. In the majority of cases the purity of the ice is really not affected, as any filth which gets on to it is melted off; but the moral effect on the customer is considerable, and it would create a more wholesome impression, and probably a greater sale, if these dirty and unnecessary practices could be avoided. There is no reason why ice should not be delivered in clean waggons fitted with insulated covers. The wastage would be considerably lessened, and the ice would be firmer and would last longer.

In sending ice by rail an insulated truck should preferably be used. The best method of packing ice in an ordinary truck is to cover the bottom of the truck with a layer of clean straw and pack the blocks of ice on this with a wooden slat between them; there should be a 6-inch or 12-inch space between the blocks and the sides and ends of the truck. The blocks should be secured in this position by timber packing, and the space between the ice and the truck should be filled with fine shavings or chopped straw, the former being much the better. Sawdust should not be used, as it pits the ice. The top of the blocks are also covered with shavings, the whole being covered with a tarpaulin.

In America it is customary with some firms to pack all ice for transport in a special straw paper. The paper is in rolls, and the ice, when packed in the car, is covered with it top and sides. It is stated that this paper only costs about eighteen-pence per truck, and that it is of considerable benefit in keeping the sawdust, shavings, etc. used as insulation off the ice.

When small orders are being fulfilled, the blocks of ice can be packed in bags, the ice being surrounded as much as possible by shavings or chopped straw.

Proper tools should be used in handling ice. These consist of breaking-out bars, lifting-tongs, saw, dagger, and axe. The use of the proper tools[1] in breaking up ice will save a lot of waste. A saw is better for this purpose than a dagger or axe, as it is not so liable to splinter the block.

[1] For complete list of ice-tools, see Catalogue of Messrs Wm. T. Wood and Company, Arlington, Mass., U.S.A.

**Selling Ice.**—Ice is always sold by weight, a cubic foot at 32° F. weighing 57½ lbs. Can ice is exceedingly convenient for distribution in this way. The cans are made to hold blocks of definite weight, running a definite number to the ton. This makes calculation easy, and weighing is frequently dispensed with. For wholesale trade this is quite satisfactory, but, in order to encourage the sale of ice, a retail trade should be carried on. Many people who will not take the trouble to send to the fishmonger's for ice will buy it regularly if it is brought to their door, and though the distribution is the most expensive item in a business of this nature, it is in many ways the most lucrative. Retail prices can be charged for ice delivered, and if the retail trade is encouraged and the wants of the householder catered for, a regular supply of ice becomes a necessity to him. When this happy state of affairs has been produced, definite rounds can be assigned to the various carts, and the daily delivery of ice throughout the summer becomes a matter of course as in an ordinary tradesman's business. The trouble and expense of collecting the accounts has been urged against house to house distribution. In the past this has been found to be a very serious matter, and a large amount of money has been lost both through bad debts and dishonest drivers. The trouble, however, has brought its own remedy, and in the United States, where the daily retail trade is very largely carried on, the great bulk of the ice is sold on what is known as the coupon system.

This system has the advantage that no haphazard cash sales are made by the drivers of the waggons, and the ice is nearly always sold cash in advance. If the customers are well known the same system is used, but credit is given. The coupon books are printed in various sizes with distinctive numbers, and contain in the aggregate coupons equal in number to 300, 500, 1000, 2000 (1 ton American), or even 10,000 lbs. of ice.[1]

The rates at which the ice is sold vary, the small coupon books being sold at double the rate of the large; this is necessary in order to cover the extra cost of distributing.

The way the system works is as follows:—

Suppose, for instance, that a consumer buys a 500-lb. coupon book and pays for this, say, 7s. 6d. On receiving the book he signs a receipt form, which is supplied with each book, and which is filed against his name at the office of the Ice Company. The actual amount of ice supplied to him daily may be only 10 lbs. or 25 lbs., or any other quantity; but whatever it is, the customer, on delivery of the ice, tears out of his book, and presents the driver with, coupons for the amount of ice received. The driver has started on his round with, say, 2 tons of ice. If it is all sold he has on his return to produce coupons for that amount. A certain definite amount of wastage is allowed him, but besides this he has to account in coupons for all the ice taken out. If at the end of the season the ice taken by the customer is less than the coupons paid for, cash equal to the value of those unused and still remaining in the book is returned to the customer, or his account is credited with their value.

The system of making the customer sign a receipt for the coupon book prevents dispute as to the number of books he has had. The amount of book-keeping and the number of bad debts is enormously reduced, and, as the drivers have not to handle cash, there is no inducement

[1] In England the coupons would probably be made out for 560, 1120, and 2240 (1 ton British) lbs. of ice.

to them to return ice as wasted that has really been sold.

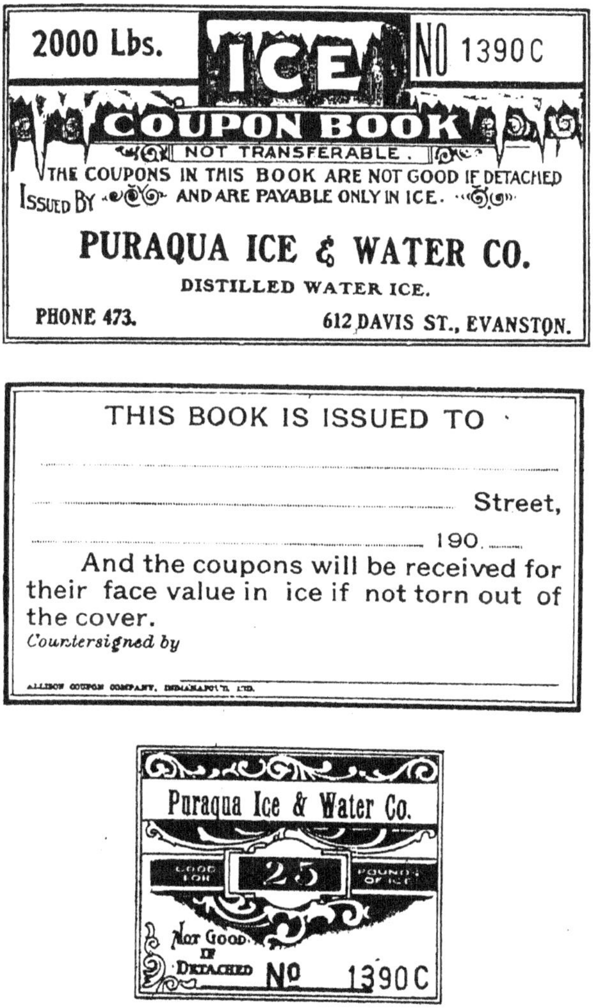
2000 Lbs. ICE NO 1390C

COUPON BOOK

NOT TRANSFERABLE.

THE COUPONS IN THIS BOOK ARE NOT GOOD IF DETACHED

ISSUED BY AND ARE PAYABLE ONLY IN ICE.

PURAQUA ICE & WATER CO.

DISTILLED WATER ICE.

PHONE 473. 612 DAVIS ST., EVANSTON.

THIS BOOK IS ISSUED TO

............................................................

........................................................ Street,

........................................................ 190....

And the coupons will be received for their face value in ice if not torn out of the cover.

*Countersigned by*

Puraqua Ice & Water Co.

GOOD FOR 25 POUNDS OF ICE

NOT GOOD IF DETACHED NO 1390C

FIG. 94.

The coupon system has been brought to a high state of perfection by the originators.[1] It has been almost universally adopted in the United States, and is preferred by the customers as well as by the manufacturers. The front and back page of, and a sample sheet from, a 2000-lb. coupon book is shown in fig. 94.

It will be observed that the books are numbered in series on both covers and coupons. By the use of these numbers it is possible to tell the exact amount of ice each customer buys, how often he takes it, and on what days. The size of the book sold is a matter to be determined by the company; if 500-lb., 1000-lb., 2000-lb. books are issued, the 500-lb. books would be numbered from 1A upwards, the 1000-lb. books 1B upwards, and so on. In this way it is possible to tell from the detached coupons what kind of book a customer has and how much he is paying for the ice. The name and address of the purchaser and the number of the book is recorded at the offices of the Ice Company when the book is sold, and no ice is delivered unless a coupon is obtained at the time of delivery. If the customer is absent, coupons should be left by the ice-safe for the amount of ice required.

The book illustrated above contains eight 25-lb. coupons, twenty-four 10-lb. coupons, and twelve 5-lb. coupons. Each value is printed on different coloured paper, so that it can be easily distinguished. As the coupons are collected, they are dropped by the driver into a small box secured by lock and key and attached to the waggon. At the end of the day the box is opened, the coupons counted, and the driver checked.

[1] Messrs the Allison Coupon Company, Indianapolis, U.S.A.

# CHAPTER XII.

## COLD STORAGE.

**General Remarks.**—The selection of a suitable site for a cold storage warehouse is a matter requiring very careful attention. As in most things, no hard and fast rule can be laid down, and the position of the site depends very much on the class of business that it is intended to carry on. If a purely warehousing business is to be followed, then the store must be within easy reach of the merchants and markets from which it is intended to draw, and to which it is desired to deliver, the produce stored. If, on the other hand, a distributing business is contemplated, the store should adjoin a railway line. In most cases railway communication will prove a convenience for coal as well as traffic, and it may be accepted as an uncontrovertible fact that the less the goods have to be moved and handled as they pass from the producer to the consumer, or in this particular case from the merchant to the cold store and back again to the market, the better for them and for those who desire to make a profit out of their storage.

Other points also must be borne in mind. Refrigerating machinery requires water and plenty of it, and it is a point of paramount importance that, if possible, this water should not have to be bought. It is not always easy to obtain a natural supply from wells, and for this reason considerable space has been devoted in Chapter VII. to the consideration of condensing water recoolers, by the use of which the same water can be used over and over again.

The suitability of the ground for receiving the foundations of the buildings and machinery is very important, and before a site is definitely selected trial holes should be sunk to ascertain the nature of the strata. A rocky foundation is, of course, best, but a subsoil of gravel or hard clay is very suitable. The ground should, if possible, be free from surface water; but if this, unfortunately, exists, it can be overcome by a damp-proof course in the concrete floor and in the walls below ground-level. Soft or unstable ground should be avoided, as piling is expensive, and concrete rafts, though a perfectly ordinary and successful method of construction, are better done without, if possible, on the score of cheapness. The question of cart space should also be considered, and it is desirable to allow draw-in space for the carts and waggons receiving or delivering goods, so that the traffic in the street or on the footpaths is not interfered with. This consideration also holds good in the case of an ice factory. It is always desirable, if possible, to purchase the freehold of the site, as leasehold properties usually mean increased expense in future years. Space should also be allowed for extension. This should be attended to at the outset and not left until afterwards, as, when extensions are in contemplation, adjoining property is apt to increase in value to a very remarkable degree.

## BUILDINGS.

The best shape for a cold store, other things being equal, is that which will give the maximum cubic capacity for the minimum wall area. This is found in a cube,[1] and therefore the more nearly square a cold store can be built, and the smaller the wall space that has to be insulated, the smaller will be the demands likely to be made upon the refrigerating machinery.

Cold storage buildings are essentially of the warehouse class. They should not be overburdened with architectural features, as these only increase the expense and add nothing to the revenue. The buildings consist usually of the engine and boiler-house and coal-store, the workshops and stores, the cold storage rooms and the offices. If ice is manufactured there is also the tank-room and the ice store. With the exception of the stores, in which the temperature has to be regulated, the buildings are quite of the ordinary type, consisting of stone or brick in mortar or cement. The cold storage rooms should be built with an outer wall of stone or brickwork, but with all the rest of the structure in timber.

If the foundations are on solid ground, ordinary footings on cement concrete are all that are necessary. If, however, the ground is shaky, piling has frequently to be resorted to. As an alternative to piling, buildings are sometimes erected on a raft composed of concrete having old railway rails buried in it. In this case the raft should have ample spread outside the outer tier of footings, or it will be apt to buckle down at the edges and crack the walls.

The engine and boiler-house should be entirely separate from the cold stores, and should have a separate entrance, not only for the hands, but for the coal and stores. The offices, as a rule, are situated on the street front, and should be so placed that the people and goods entering or leaving the premises can be observed from the windows. All

[1] See note, page 294.

these points are common to ordinary warehouses and business premises; the construction of the engine and boiler-houses, chimney-shaft, etc., and of the offices, are well-known engineering and architectural matters. It is the construction of the cold stores themselves that is special, and it is in this particular that considerable experience and skill are required.

**Materials Used.**—It has been remarked that the cold storage rooms should be constructed of timber. The reason for this is, of course, that, in itself, timber is a poor conductor of heat. Were the internal structure built of steel and concrete with iron columns, the greater conductivity of the iron or steel would render it much more difficult to maintain a low temperature. This difficulty is particularly marked where different classes of goods are stored, each of which requires a specific and unvarying temperature, and it has been found almost impossible to keep a definite temperature in rooms where any iron has been used in the construction. If iron columns are used they must be encased in insulating material—this, of course, costs money and takes up space; with timber it is quite unnecessary.

Another point in favour of the employment of timber is that in case of fire no damage is done to the brick walls through the warping and twisting of beams, such as always takes place when metal is used.

Hennibique ferro-concrete has recently been used for the construction of cold storage premises. This construction consists of concrete containing iron or steel rods, which are tied together. Experiments have been made which prove that concrete is but a poor conductor, but, quite apart from the question of cost and time of construction, it is difficult to see what advantages from a cold storage point of view this method of construction has to offer over brickwork and timber.

It is generally necessary, in order to conform with the various building laws of the country, to construct the outer shell of the cold storage building of brickwork and to roof it with slates or tiles. In Colonies the cold rooms are built entirely of timber, the walls being constructed as shown in fig. 73 (page 226); the wall itself contains the filling material and forms the insulation. In this case the weather-boarding is painted white, and naturally throws off the sun's rays.

The same system should be followed with brick buildings, and although white glazed bricks are out of the question on account of expense, the exterior brickwork should be kept whitewashed, and should be coated on its inner face with a coat of cement or pitch, as pointed out in Chapter VIII.

## Classes of Cold Stores.

Cold stores can be divided into two general classes :—

1. Those in which only one general class of goods is stored, a uniform temperature being carried throughout; and
2. Those in which various classes of goods are stored, different temperatures being maintained in each room.

**Class I.**—The first cold stores, being used for frozen meat, were built in this way, a uniform temperature of 22° being maintained throughout the building.

In a store of this kind the construction is wonderfully simple, and consists of a brick casing completely lined with a layer of insulating material, the structure of wood supporting the various floors being built inside this and to as great an extent as possible independent of it.

In some stores belonging to this class which have been erected, and which are used principally for meat, the only entrance is from the top, and all goods that enter or leave the stores have to be hoisted up to the level of the top floor and lowered into the store. On leaving the store they have to be again hoisted up and lowered to the ground-level. In some cases the greater portion of the top floor is devoted to distributing and sorting the various brands of mutton and beef. The whole of the internal structure is of wood, and the floors are laid with open joints, so that there is a free circulation of air from the top to the bottom. This method of construction undoubtedly makes a smaller demand on the refrigerating machinery than any other, but this advantage is to a certain extent counterbalanced by the additional cost, both in labour and coal, of handling. There is little advantage in saving steam in the refrigerating machines if it has to be spent in the hydraulic pumps or electric plant. Even where only one class of goods is stored it is now considered better practice to lightly insulate the floors and subdivide the building into independent floors and rooms, each room being in communication with a corridor, and each corridor with one or more lifts or hoists. By this means the goods can be taken in at ground-level, and need only be lifted to the level of the particular floor on which they are to be stored. If the cold storage room doors and passages are properly constructed, as pointed out on page 335, and fitted with port-hole doors for the passage of small articles, there is very little tendency for the cold air to flow out, and, even if it does, the cost of removing the heat introduced in this way is counterbalanced by the lessened cost of lifting and handling. The cooling of a store having only one temperature is also an exceedingly simple matter, and can be done by either of the three systems dealt with subsequently. Although some stores of this class have used brine circulation, air blast or direct expansion are the most common. Direct expansion is the most economical in working.

**Class II.**—It is in this class of cold store that great forethought and skill have to be shown, and the design becomes much more complex. The store has to be divided up into a number of different rooms, each insulated from the other more or less heavily according to the temperature difference. As a general rule, however, it is not necessary to insulate the floors and partitions, as they will answer the purpose very well if they are made air-tight with paper. Each of the rooms has to be in easy communication with the lifts and hoists, and each has to be internally fitted to suit the class of goods intended to occupy it. Not only has each room to be refrigerated in summer, but it has to be warmed up in winter,[1] for with certain goods it is fatal to vary the temperature more than one or two degrees each way. In fact, so important is this point as affecting the quality of the goods, that, in America, Cold Storage Companies have often to give an undertaking before the goods are left that the temperature will not be allowed to vary outside the prescribed limits. The rooms have to be ventilated, and, in order to avoid shrinkage, the humidity has to be kept at a definite point.

In general construction the rooms follow the timber construction mentioned in connection with the former class. Each floor is separated from the other, and bulkheads or partitions are placed where necessary in order to subdivide the rooms.

[1] This, thanks to our climate, hardly applies to English practice, but it is common in the States.

In London the Building Act forbids any single space having a greater cubic capacity than 250,000 cubic feet, and to meet this large stores have to be divided up by brick walls and fireproof doors. If it were not for this it would be better, from the cold storage point of view, to have no brick walls except the external ones.

Fig. 95 (page 321) shows a section through part of a cold storage building of this class; this is intended only to illustrate the most suitable method of construction, and does not pretend to go into exact detail. It will be noticed that the brick wall of the building proper is entirely distinct from the rest of the building, and is divided from it by the insulation. The internal portion of the structure is entirely of timber. Except here and there, where they are required to act as ties, the various floor beams do not run through to the wall, but are carried on timber columns direct from the floor. By this means the insulation is kept unbroken.

The loading in and out platforms and the receiving floor should be at tail-board level, and from this platform the goods should be trucked to the lifts or hoists and thence to their proper rooms. Receiving, checking, and weighing offices should also be provided on the receiving platform, so that all goods received and taken out can be weighed and observed.

The rooms in which goods are placed, which are only wanted to be kept in storage for a short time, should be situated on the ground floor. If there is an egg-room it should also be on this floor, so as to save handling. Throughout the rest of the building the rooms should be arranged as is found most convenient, the general principle being observed that the rooms carrying a low temperature should be kept together near the centre of the building and apart from the rooms carrying a higher one, which should be on the outside. By introducing several extra walls between the outside atmosphere and the freezing chamber, and by reducing the temperature difference between adjoining rooms, this arrangement enables lighter insulation to be used in the partitions.

The rooms should, in addition, be so grouped that different classes of goods of the same general character can be stored together or near each other—thus meat and poultry together, fruit together, vegetables together. Cheese, butter, and other sensitive goods should, of course, be kept quite separate.

With stores of this sort all pipes conveying ammonia, brine, or steam should run along the corridors. All expansion valves and regulating cocks should be placed in the corridors within easy reach of the floor, and no pipes, except the actual cooling coils, should be placed in or allowed to pass through the various rooms.

## Methods of Cooling.

There are in general use three principal methods of refrigerating cold stores. These are known as:—

1. Direct expansion.
2. Brine circulation.
3. Cold air blast.

A rule ordinarily followed is for direct expansion or air blast to be used for temperatures below freezing, and brine circulation[1] or air blast for temperatures above freezing.

**Direct Expansion.**—The direct expansion system consists

[1] In this case care must be taken to arrange the brine pipes so that any drip can be caught and drained away by gutters as shown in fig. 96.

merely of a coil of pipes placed on the ceiling or on the walls of the cold rooms, in which the gas expands, absorbs heat, and produces the refrigerating effect. The gas is expanded directly into the coils, and is drawn through them by the action of the compressors. The pressure in the coils varies from 10 to 30 lbs. according to the temperature required in the rooms; 25 lbs. is a very usual pressure, and corresponds to a temperature of about 22° F. As the gas begins to expand in the coils, the temperature of the pipes is reduced and the moisture from the air is deposited upon them in the form of snow. This, of course, dries the air. The snow is scraped off the pipes from time to time. Some makers use radiating discs upon the pipes; these, however, are of little use, and are very much in the way when the pipes have to be cleaned.

The expansion cocks should be situated in the passage just outside the room in which the coils are placed, or in some other convenient position. Direct expansion is by far the most simple method of cooling, and, owing to the higher back pressure that can be carried, it is the most economical.

As the temperature depends on the pressure, direct expansion does not easily lend itself to large variations in temperature. If the back pressure of the gas returning from the various circuits differs largely, trouble is met with and uneven working results. This can be got over to a large extent by having three or four coils in the same room, the temperature being regulated by turning one or more off or on and increasing or decreasing the cooling surface. The back pressure in this case remains the same throughout the building, but the regulation is not so easy as with brine circulation. Direct expansion

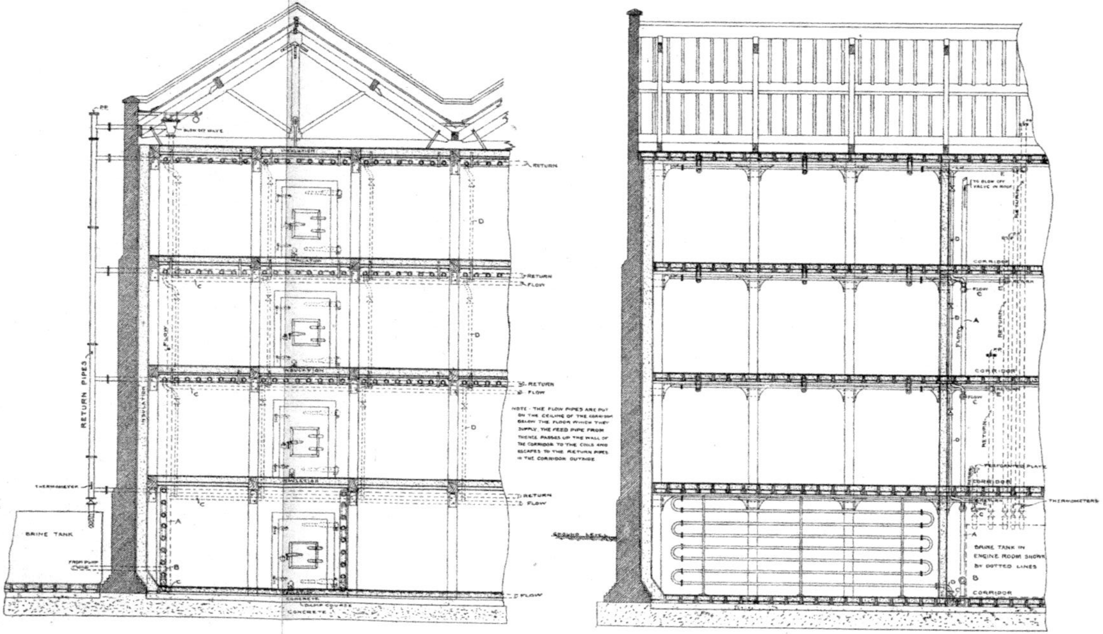


TRANSVERSE SECTION
THROUGH COLD ROOMS

LONGITUDINAL SECTION
THROUGH COLD ROOMS

FIG. 95.

*To face page 321.*

is most suitable for stores in Class I., in which only one temperature is maintained.

**Brine Circulation.**—In this system the ammonia, or $CO_2$, is expanded in coils and utilized to cool down brine, which is circulated through pipes placed in the various cold stores.

The brine is drawn from the cooler and is kept under a pressure, depending on the circumstances of the building, of about 10 lbs. per square inch by a pump. Under this pressure it is circulated through coils placed in the various rooms and returned to the cooler, where it is again cooled. As the brine is all of the same temperature when it leaves the cooling tank, it is obvious that in order to vary the temperature in any room it is only necessary to regulate the quantity of brine flowing through[1] the coils placed in it. Nothing could be more simple or more effective.

Great care has to be exercised in placing and installing a brine system, as, in addition to the refrigerating, the hydraulic side of the question has to receive attention.

Fig. 95 shows a longitudinal and transverse section through a portion of a cold store fitted with brine circulation. It will be noticed that the brine flow and return headers are situated in the corridor, the flow for one floor being in the corridor below the rooms, the return being in the corridor on the same level as the rooms. By this arrangement the brine is always gently flowing from a lower level to a higher level, and there is no tendency for air to get locked in the pipes.

[1] In some stores carrying different temperatures, a brine mixing tank is introduced and the cold brine from the cooler is mixed with a warmer return brine. The mixture is used for the higher temperatures, thus ensuring an even temperature in all parts of the room. The system requires extra pumps and cross connections.

The brine-coils can be placed on the walls, as shown in fig. 96,[1] or on the ceiling, or both; whereas, if the flow and return pipes were both in the same corridor, pipes could, to give good results, only be placed on the ceiling.

The flow of brine through each room can be regulated by the valves, and a whole coil can be shut out if necessary. The general system of the arrangement of brine piping is as follows:—The cold brine from the brine cooler enters the stand-pipe *A* at *B*. This stand-pipe runs right up to the roof of the building, and is fitted at the top with a dead-weight, blow-through valve, loaded to lift at, say, 12 lbs. per square inch. The overflow from the valve passes back into the brine-tank through a separate pipe. This valve is provided to guard against accident, should all the deliveries be shut off at any one time while the pumps were running; in ordinary work it is seldom or never in action.

From the stand-pipe the flow to each floor passes off into the headers *C*, which run along the corridors; that for the ground floor rooms is encased in the floor as shown. The cold brine leaves the headers through the pipe *D* outside the cold room, passes through the wall, through the grid of pipes on the ceiling or wall, and back to the return header *E* placed in the corridor above. The wall and ceiling coils should be on separate circuits controlled by separate valves, so that they can be turned on or off, or regulated independently of the others. Where the pipes from the various circuits enter the return header, they are fitted with a thermometer, so that the temperature of the brine can be read at any time. The return header runs along the passage and discharges into the brine-tank as shown. Each header

[1] Reproduced from *Ice and Refrigeration*, Chicago.

FIG. 96.

*To face page 322.*

is at this point provided with a thermometer, by means of which the engine-driver can see at a glance the temperature of the brine from the various floors.

It will be observed that each floor is one main circuit, and that the various rooms form branch circuits from this. The return headers passing down to the brine-tank must have a perforated flange at the top of the vertical portion, or a small pipe running up above the level of the highest floor, in order to break the vacuum and prevent the brine siphoning out of the pipes. The lowest point of the system should be connected to the suction of the brine-pump through a small pipe, so that the brine-pipes can be pumped out at any time.

As the resistance to the flow of brine through the lower coils is less than that through the upper, there is nothing to prevent the brine-pipes on the upper floors emptying through those on the lower when the pump is stopped. In the majority of plants this difficulty is overcome by fitting each main return pipe with a valve, and closing it, and the valve by the pump, when the latter is stopped. The flow of brine through the various rooms is rendered even by merely regulating the valves on the various floors. In some designs, however, the return pipes from each flow are taken to the level of the top floor and then returned to the brine-tank. The former system is that most commonly in use, and it saves power, as the brine does not all have to be pumped to the highest level.

In other designs each separate circuit throughout the building returns to the brine-tank, where it is labelled and fitted with a thermometer and valve. The flow of brine through any circuit is controlled from this point more or less, being passed according to the reading of the thermometer. Before the brine-pump is stopped these valves are all shut off, and the brine is retained in the coils. This arrangement is very simple and effective, but, where a variety of temperatures have to be carried, is not so convenient as the system first described.

In order to secure an even flow through the various pipes, the coils should all be of approximately the same length. Too large circuits should be avoided, and with $1\frac{1}{4}$-inch pipes 400 feet is about the maximum advisable. With 2-inch pipes this may be increased to 1000 feet. Practice, however, varies, and the circumstances of each case must determine the lengths of the various circuits.

Under normal circumstances the brine will gain some 6 to 8° F. in temperature in traversing the coils.

The best spacing for the pipes is in grids at not less than 9-inch centres. Where chilled beef is being hung the pipes are frequently nested between the rows of carcasses. Brine walls, somewhat similar in construction to those used in the cell and plate ice-making plants, are frequently used for this purpose.

In cases where two or more floors are all kept at the same temperature and no insulation is used in the floors, the brine-pipes can be conveniently nested between the main timbers of the floor; in this case, as before, the brine enters at the lower level and flows away to the return header at the higher level.

**Brine-Tank and Brine-Pumps.**—The brine-tanks containing the evaporating coils can be of the ordinary oblong pattern as in ice-making, but, in order to save room, it is most usual to make them more after the pattern of the submerged condenser, the brine surrounding the evaporating coils, which are contained in a vertical circular wrought-iron tank.

Brine-pumps need not be of any special description, but they should have their working parts of gun metal, and should be provided with an air vessel on both suction and delivery, as well as a bye-pass. Fig. 97 [1] shows an arrangement of brine-tank pumps and brine coolers, and will be sufficiently clear without any other explanation.

The brine coolers are of the Pontifex type, and vary from other kinds in that the brine is passed through coils

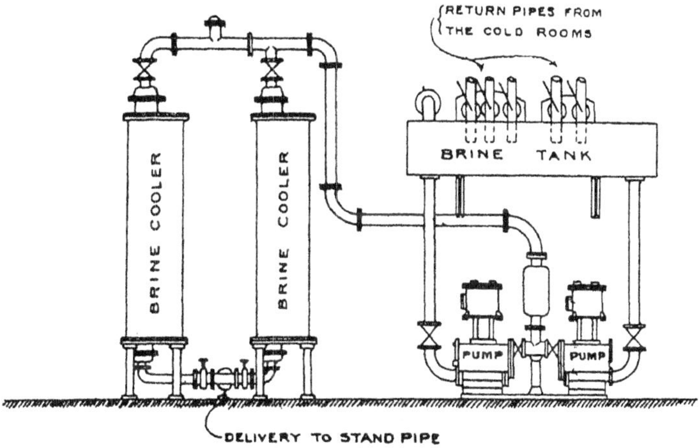


Fig. 97.

contained in the casing, the liquid ammonia being round them. The cooler is illustrated in fig. 98. It is important that all brine-tanks and coolers should be efficiently insulated, and they are best provided in duplicate as in fig 97, one being used as stand-by.

**Air Blast.**—In the air-blast system there are no pipes in the cold rooms. Air is blown over a battery of pipes in which ammonia is expanding and over which brine is

[1] Messrs the Haslam Foundry & Engineering Company, Ltd., Derby.

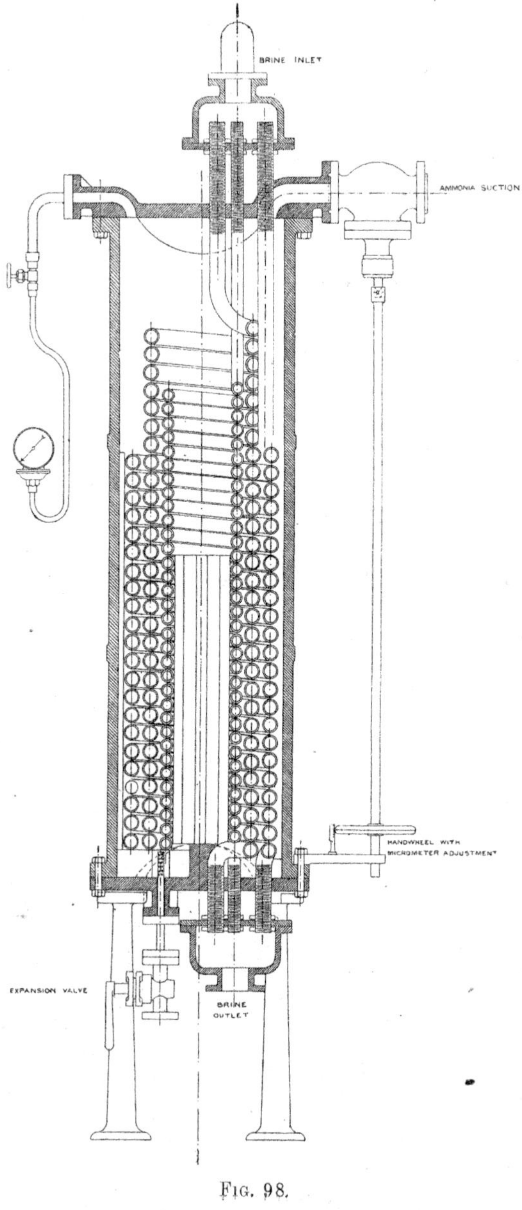
BRINE INLET
AMMONIA SUCTION
HANDWHEEL WITH
MICROMETER ADJUSTMENT
EXPANSION VALVE
BRINE
OUTLET

FIG. 98.

kept trickling. The battery is placed by itself usually at the top of the building in an insulated box or trunk. The air is drawn from the stores by a powerful fan and forced over the coils. At the same time as it is cooled down it is robbed of its moisture. Instead of pipes some makers use metal surfaces kept wet by refrigerated brine.

In some cases brine is dispensed with and the air is blown over dry coils. The moisture in the air deposits on the coils in the form of snow, and has to be cleared off from time to time. Brine batteries are also made of a series of corrugated galvanized iron plates hung close to each other; over these, cold brine is kept circulating, and through them the cold air is blown. Owing to the corrugation of the iron, the air is brought intimately into contact with the cold brine and is chilled rapidly and effectually.

By these methods of cooling the air is rendered very dry; any moisture it originally contained, or which it may have picked up from the goods in the store, is deposited on the coils of the battery and absorbed by the brine. The air trunks are arranged with one main delivery from the cooler and one return to the cooler. Each cold room has its branch delivery and return trunks. These are fitted with slides placed at intervals along their length to regulate the amount of air that goes into each room. Inside the rooms the delivery main usually runs along the top of one side and the suction along the top of the other side of the room. The cold air enters through the delivery trunks, and, being colder than the rest of the air in the room, falls to the bottom. As it absorbs heat from the goods or the room, it gets warmer, rises, and is drawn to the cooler by the suction main. At the cooler it is deprived of its heat and again circulated.

Sluices are sometimes made in the mains, so that fresh air can be introduced and the return air allowed to go to waste, but, as a rule, sufficient ventilation is considered to be obtained by the opening and shutting of doors in daily working.

The air-blast system is the cheapest to instal, and is very commonly used. It has, however, two very serious drawbacks. These are:—

1. The air is rendered exceedingly dry, and has been found to cause considerable loss through shrinkage in weight of the goods stored.[1]
2. As the air is drawn from all the rooms and blown through a common battery, the air from each room is mixed up with that from its neighbour.

It is easy to see that if butter is stored in one room and, say, onions or apples in another, the air drawn from these rooms will be contaminated with the flavour of the goods stored, and the flavour of the apples, etc. will be communicated to the butter.

It is out of the question, on the score of expense and the space occupied, to have a separate battery for each room, and except in stores of even temperature, and only containing one class of goods, the air-blast system is not advantageous. It is, further, difficult when goods are constantly being taken in or out to regulate with any exactness or ease the temperature of a room cooled by air blast. The suction and delivery trunks are provided with slides, and these can be regulated until the desired temperature is obtained. The air, however, will take the path of least resistance, and it is plain that unless

[1] This is most serious at temperatures below freezing ; above 32° Fahr. the air can be kept more humid.

the speed of the fan is altered, the regulation of the slides in any one room will alter the amount of air passing through the others. This alteration is not so great when the regulation is done on the suction slides, but even then it is noticeable, and, where definite temperatures are necessary, the system introduces considerable difficulties.

The most common method of fitting the air trunks in a room is shown in fig. 99. It will be noticed that the delivery is in one top corner of the room and the suction

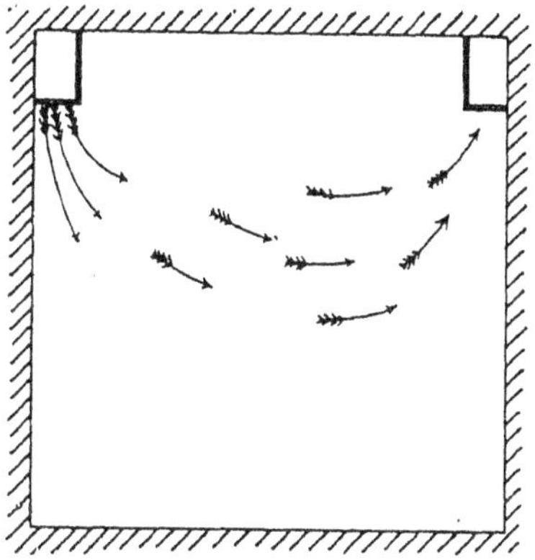

Fig. 99.

in the other. With this arrangement the air will take the course shown by arrows. The arrangement is faulty for this reason:—the air is travelling at a velocity, and will naturally follow the course of least resistance. It will descend from the delivery duct into the cold room, where it will strike against the top of the goods stored and pass off to the return duct. It must be remembered that under usual working conditions there are not many degrees difference between the incoming and outgoing air, and that circulation caused by the greater specific gravity of the former cannot be relied on to any great extent. The air in the cold room will be cold, but plainly there will be little or no circulation of the air on or near the floor of the room, and the air in between the goods will practically remain dead and unchanged.

This point is strongly controverted by some authorities, who maintain that the air *is* changed. The importance of the point will be realized when it is remembered that almost all goods in cold storage except meat will 'gas'—that is to say, they will give off large quantities of gas, usually carbonic acid. With fruit and eggs this is particularly the case, Bartlett or Williams pears and lemons being, with fruit, probably the most notable examples. If the fruit is to be kept in sound condition this gas must be removed as soon as it is formed. If it is allowed to remain the fruit will become mouldy and go bad. As the carbonic acid is heavier than air, it will naturally tend to lie on the floor of the cold room and, with the above system of air circulation, will not be disturbed. That this is the actual state of affairs cannot be denied, and it has been stated that a lantern which will burn brightly in the upper portion of a fruit-room cooled in this manner will go out if placed on the floor The remedy for this faulty circulation is to be found in the rearrangement of the air ducts. There are several positions in which they can be placed; one of the most suitable is shown in fig. 100, where it will be observed that the cold air is introduced through a duct in the ceiling and the warm air is taken away by a duct in the floor. The quantity of air entering the room is regulated by a sluice in each duct. The suction duct passes under the floor, and the air leaves the room through a number of perforations in the wooden cover.

It is immaterial whether goods are placed on top of the duct or not, as there will always be some holes uncovered and a way open for the escape of the air from the room.

**Combined Brine Circulation and Air Blast.**—Though this arrangement of air ducts gets over the bad circulation of the air in the chambers and removes the foul gases given off by the goods, it does not affect the two principal defects which are inherent to the air-blast system. As usual the remedy lies in combination and compromise,

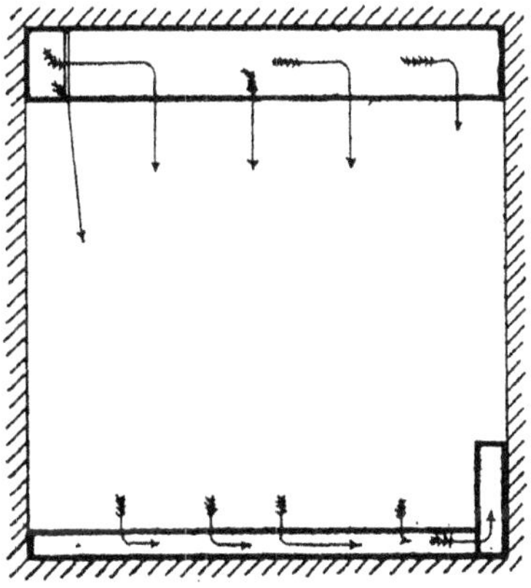

FIG. 100.

and is found in a combination of brine circulation and air blast—the brine to cool the rooms and to afford an easy means of regulating the temperatures; and an auxiliary battery for ventilation. The battery is connected to each room through delivery ducts only, and in many cases need not be kept constantly in use, but only worked from time to time when ventilation is required. The cold air from it enters the room at the top and leaves it at the bottom through an escape duct, which communicates with a central shaft connected with the atmosphere.

Fresh air from the outside atmosphere is drawn in by the fan and blown over the battery, and the same air is never used twice. In fruit-rooms, where gas accumulates rapidly, it is best to keep up a constant gentle circulation of fresh air. As the fresh air is cooled before it enters the room, there is no extra work thrown on to the brine-pipes, and their regulation need not be altered. As the cooling is done by brine-pipes, there is no tendency for the air in the rooms to become too dry, and, as the foul air is sent to waste, there is no possibility of the goods in one room communicating their characteristic odour to those in another portion of the building. If the air ducts are put in when the building is being erected, the only extra cost is in the battery, which is not in itself an expensive item.

**Natural Ice Cold Storage.**—Another system of cooling cold stores in which, however, no machinery is employed, has been introduced in America. The system relies on an abundant and cheap supply of natural ice, and the cooling is effected by the frigorific effect of a mixture of ice and salt. The rooms are cooled by brine-pipes on the ceiling or walls, or by an air battery as the case may be. The brine in the pipes does not come into direct contact with the ice and salt, but is circulated through a coil of pipes placed in a tank at a high level which contains the frigorific mixture. This coil of pipes is connected to the brine-coils in the rooms at a lower level, and as the brine in the coils in the tank becomes chilled by the salt and ice, its lower specific gravity causes it to fall to the cooling coils below. Here it absorbs heat from the rooms; becomes warmed and rises to the first tank, to be again chilled by the ice and salt. The circulation is practically the same as the circulation of water in a kitchen boiler. Obviously the system can only be economically used where natural ice is plentiful, cheap, and easily harvested; and though exceedingly ingenious and of considerable use in Canada and the northern states of U.S.A., where these conditions exist, it could not, of course, compete in England with mechanical methods.

## DETAILS.

**Details of Construction.**—The cold storage rooms are usually from 7 feet 6 inches to 8 feet high, and there should be not less than six feet clear between the pipes on the ceiling and the floor. The floors can be of either pitch-pine or iron bark gum; both make good wearing surfaces.

If loose articles difficult to stack are being stored, valuable space can often be saved by subdividing the rooms into bins or divisions. These should not be permanent fixtures, but should be movable and capable of being placed in any desired position. A satisfactory way of doing this is to make a hurdle in which the bottoms of the two uprights fit into a recess in the floor, the tops being held in position by two short pieces of angle iron secured to the ceiling.[1]

The size of the rooms varies with the class of goods stored and the business done. It is more easy to maintain a regular temperature in large rooms than in small ones, but, on the other hand, if there is a great variety in the different goods to be stored and the temperature to be kept, smaller rooms save space. It may happen that in a store containing only large rooms a small quantity of goods requiring a certain temperature may

[1] This design was introduced at the Victoria Docks by Mr Donaldson.

arrive. On the score of policy it is undesirable to decline to store these, and there is no alternative but to set aside a large room for their reception. Again, it is frequently advisable permanently to hire out rooms to certain merchants, and in this case a variation in the size of the rooms is desirable. In a properly designed store there is no difficulty in arranging the rooms so that they can at any time be subdivided as required by lightly insulated or air-tight partitions. This is particularly easy where brine-pipes are arranged in bays on the ceiling, each bay being controlled by its own valve placed in the passage outside.

**Painting, Varnishing, and Limewhiting.**—There is considerable difference of opinion as to whether the interior of cold stores should be painted, varnished, or limewhited. In considering this question, it must be remembered that the property of a body for absorbing and transmitting heat varies, *inter alia*, in direct proportion to its density and colour. If the inside timber is painted, the paint fills up the pores of the wood, makes it denser, and, being of a metallic nature, assists the conduction of heat; if, in addition, it is of a dark colour, it assists the radiation of heat.

Varnish puts a skin on the surface of the woodwork which renders it air-tight; it is not of a metallic nature, it does not absorb smell or moisture or allow the timber to do so, and it always looks clean. In these respects it is very satisfactory.

Limewhite is neither damp- nor moisture-proof, but it is cheap and easily put on; it is, furthermore, a disinfectant, and being white it does not radiate heat. A well-limewhited room looks clean and is easily washed from time to time. This is especially valuable when a room, which has been constantly used for the storage of one class of goods having a characteristic odour, is required for the storage of other goods of a more delicate nature. Before the room can be safely used it has to be carefully cleaned out and deprived of any odour that may cling to it by being cleansed by washing with a strong lime solution and several washings with water containing carbonate of soda.

**Lifts and Hoists.**—A lift working between the various floors of a cold store will act very much like a pump, and will cause a constant circulation of air. In order to prevent this and to keep the air in as still a condition as possible, the lift-ways should be quite encased in a lightly insulated casing. Proper doors should be provided at each floor and only opened when the lift is in position. The air in the corridors outside the cold rooms is thus kept dead and at an even temperature. The corridor forms a lobby or air lock to each set of rooms, and there is a lessened tendency for the cold air to flow out of the cold rooms when the doors are opened to admit or withdraw goods. With the same object doors should also be provided on the stairways at each landing.

The lifts are best placed in a recess; this keeps the corridor clear and allows sliding doors to be used in front of the lift openings.

For all-round work lifts are the most suitable, as the goods can be sent up in trucks, wheeled direct into the cold store, and placed in position with the minimum of handling. It is a great mistake to consider that because meat is frozen hard it can be knocked about to any extent without suffering depreciation; this is not the case, and a carcass that has been severely handled in loading or unloading never looks well when it is thawed out and exposed for sale.

The lifts can be either of the direct-acting type or of the geared type. They can also be either electric or hydraulic; the latter are well known and the former have already been described. Hoists are different and are not so well known. A very handy form of endless chain hoist designed for handling carcasses of mutton and very useful under certain circumstances is shown in fig. 101.[1] A hoist somewhat similar can be used for handling tubs of butter, barrels of apples, or cheese. The hoist is worked by an electric motor, and the speed can be varied through wide limits, a reduced speed being necessary if fragile goods are being handled. The motor is started and stopped by solinoid control, and can be put in or out of operation from any point or from the tally clerk's office. The simplicity of the necessary lifts and hoists is one great advantage of the subdivided cold store with a bottom entrance. With the entrance from the top much more elaborate and costly devices are required.

**Doors.**—A cold storage door should be air-tight when shut, efficiently insulated, and easily opened or closed. Frequently hatchways are provided either in the main door or in the corridors for the passage of small articles. When placed in the main door they interfere rather with the proper bracing, but they are exceedingly convenient and minimize labour and the escape of cold air.

The doors should have a flush sill so as to allow trucks to pass directly into the room, and, in order to avoid any leakage, should have a canvas or felt cushion on the underside. Careful fitting and good insulation is all that is required.

Mr Stevens[2] insulates his doors with alternate air

[1] Messrs J. C. Childs & Company, Ltd., London.

[2] Mr B. A. Stevens, Toledo, Ohio, U.S.A.

FIG. 101.

*To face page 336.*

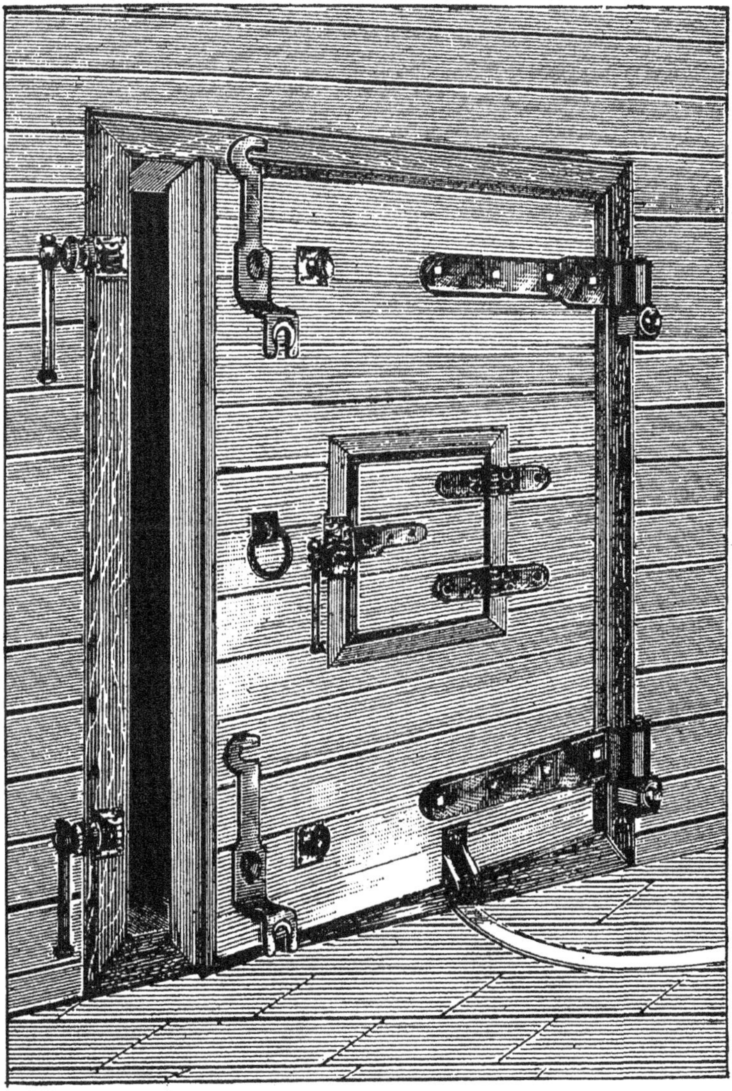

Fig. 102.

spaces, timber and waterproof paper, and has had considerable success with them. The doors are panelled and braced, and are not usually made with a hatchway. The hinges and latch are of special design, a feature of the door being the provision of tension latches on both the latch and hinge side, by means of which the door can be screwed up tightly when shut.

Another make of door is illustrated in fig. 102. This is the type commonly found in this country, and is shown fitted with Taylor's patent fastenings.[1] The illustration shows a raised sill, but this can be made flush if necessary. A hatchway is shown in dotted lines. The weight of these doors being considerable, it is always advisable from every point of view to provide them with a wheel and track as shown. It saves the door frame and hinges. and is of considerable assistance in easing the handling. All ironwork should be galvanized.

Another door fastener is shown in fig. 103.[2] It is a combined bolt-locking bar and tightener and is very simple to fix and easy to operate. The door can be easily locked and the bolt secured in position by a padlock passed through the hole shown. There are several other designs of fasteners current, the object in each case being to keep the door air-tight by forcing it up to the jambs.

**Power required for Cold Storage.**—As there is considerable doubt among the non-technical as to the power of refrigerating machines and the amount of space that can be efficiently cooled for a given horse-power, the following table may be of interest[3]:—

[1] Mr John Straiton, Bootle, Liverpool.

[2] Messrs Taylor & Kensitt, Tooting Junction, London.

[3] Messrs L. Sterne & Company, Glasgow. The power and capacity given in this table are for the speeds given, *viz.*, 36 revolutions per minute. In

| | | | | | | | | | | |
|---|---|---|---|---|---|---|---|---|---|---|
| No. of sheep, 70 lbs. each, reduced from 98° F. to 20° F. in 24 hours. | 64 | 128 | 257 | 580 | 773 | 1,160 | 2,062 | 3,223 | 4,834 | 7,091 |
| No. of sheep, 70 lbs. each, reduced from 98° F. to 32° F. in 24 hours. | 122 | 244 | 488 | 1,098 | 1,464 | 2,197 | 3,906 | 6,103 | 9,155 | 13,428 |
| No. of hogs, 200 lbs. each, reduced from 98° F. to 32° F. in 24 hours. | 42 | 85 | 170 | 384 | 512 | 769 | 1,367 | 2,136 | 3,204 | 4,700 |
| No. of oxen, 1000 lbs. each, reduced from 98° F. to 32° F. in 24 hours. | 8 | 17 | 34 | 76 | 102 | 153 | 273 | 427 | 640 | 940 |
| Well water required per minute in gallons at 45° F. | 3 | 6 | 10 | 20 | 28 | 40 | 70 | 100 | 150 | 220 |
| Horse-power required at 36 revs. 20 lbs. back pressure, 160 lbs. direct pressure. | 3 | 5 | 10 | 22 | 28 | 44 | 74 | 120 | 180 | 250 |
| Capacity in thermal units absorbed in 24 hours. | 564,000 | 1,128,000 | 2,256,000 | 5,076,000 | 6,768,000 | 10,152,000 | 18,048,000 | 28,200,000 | 42,300,000 | 62,040,000 |
| Capacity in cub. ft. of space alone, which machine is capable of refrigerating. | 16,000 | 32,000 | 74,000 | 176,000 | 228,000 | 346,000 | 600,000 | 948,000 | 1,422,000 | 2,000,000 |
| Capacity in tons of ice melted in 24 hours. | 2 | 4 | 8 | 18 | 24 | 36 | 64 | 100 | 150 | 220 |
| Capacity in tons of ice manufactured in 24 hours. | 1 | 2 | 4 | 10 | 13 | 20 | 40 | 60 | 90 | 130 |
| Single- or double-acting. | Single | Double | do | do | do | do | do | do | do | do |
| No. of compressors. | 1 | 1 | 2 | 2 | 2 | 2 | 2 | 2 | 2 | 2 |
| Size of compressors. | 6 × 10 | 6 × 10 | 6 × 12 | 8 × 16 | 9 × 16 | 10 × 20 | 12 × 24 | 14 × 28 | 16 × 32 | 18 × 36 |

**Small Cold Stores.**—A large cold store is only a profitable property on regular trade routes or in populous centres, and, as food goes bad irrespective of locality, its purveyor is frequently obliged to erect a small store of his own, in which to keep his perishable produce pending its sale. The value of such a store can only be measured by the money it saves its owner by preventing

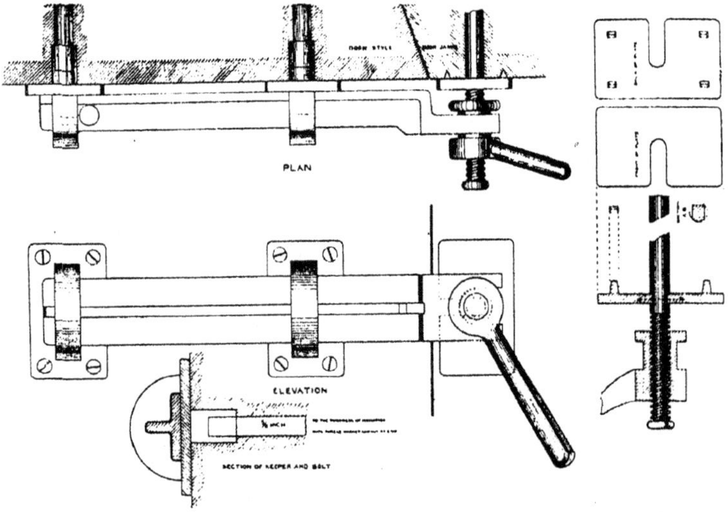


FIG. 103.

the loss and depreciation of his goods, and is an adjunct to his business in much the same way as is his shop. The advantages of cold storage from this point of view have been admirably set forth by Mr Brydon Hogg,[1] and it is unnecessary to refer to them here.

practice the first three sizes are run at 60 revolutions, the next two at from 50 to 60 revolutions, and the rest at 50 revolutions per minute.

[1] *Proceedings of the Cold Storage and Ice Association*, vol. ii. No. 1.

The stores are of two types, *viz.*, those cooled by ice and those cooled by machinery. The decision as to which type to employ is governed very largely by the size of the store that is wanted and by the cheapness of the ice obtainable. For stores of and up to 200 cubic feet, ice, if it is obtainable at all reasonably, is the better. For anything larger a small machine driven by a gas-engine is very suitable. In both cases the same principle holds good, namcly, the constant circulation of cold dry air, for

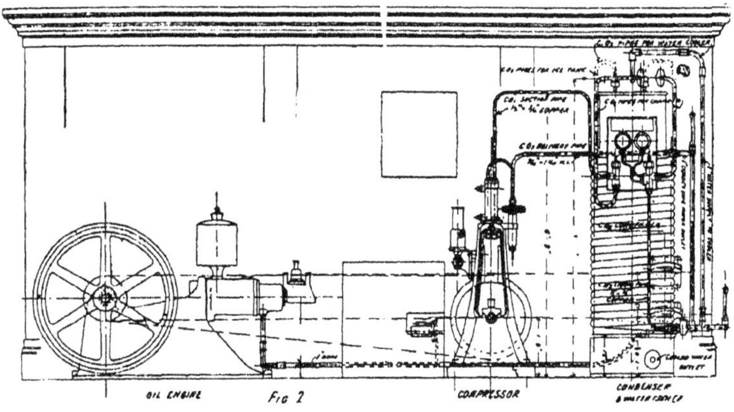


Fig. 104.

produce preserved in a dry even temperature is in the best possible position to resist decay. In Canada and the United States, whore natural ice is abundant and cheap, small cold stores are built on what is known as the Hanrahan systom.

Theso stores consist of two chambers, the ice being placod in one and the produce in the other. The ice-room measures about 13 feet by 14 feot by 20 feet high, and the store room 13 feet by 10 feet by 8 feet 6 inches high. These stores are, of course, built out of doors and can be very cheaply constructed. In England the store cooled by ice is usually placed indoors and is practically a large ice-box, this being the most suitable for small businesses. The refrigerator, however large or small, should be specially constructed by, or obtained from, people who understand the business and the circulation of air, and should preferably be built with the opening in the top. If a large business is carried on and ice is expensive

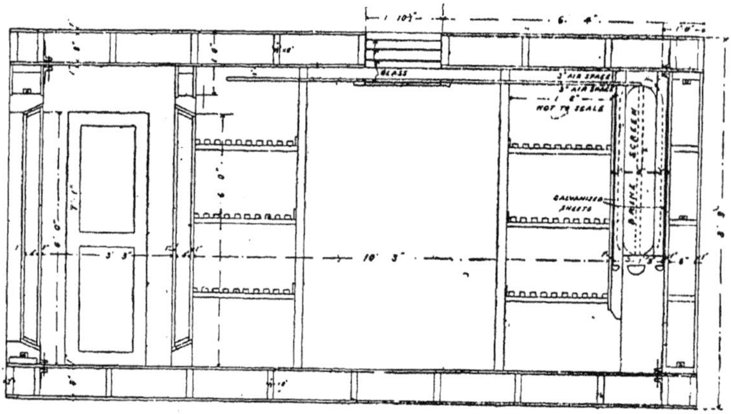

FIG. 105.

and difficult to obtain, the room should be kept cool by machinery. Small machines can now be obtained cheap in cost, self-contained in space, and simple to run. Gas or electric power is almost always at hand, or, if absent, steam or oil can be used. These small refrigerating machines are exceedingly useful in the country and for use in connection with country houses and dairies. The same machine can be used both for cooling the chamber and for making ice. In 1901 The Royal Agricultural Society offered a prize for a small refrigerating plant suitable for a dairy. There was only one competitor,[1] who exhibited a machine working in connection with a cold chamber. This chamber and plant, which is admirable for its purpose, is shown in figs. 104, 105, and 106 reproduced from vol. 62 of the Journal of the Society. The room is insulated with silicate cotton and is provided with double doors

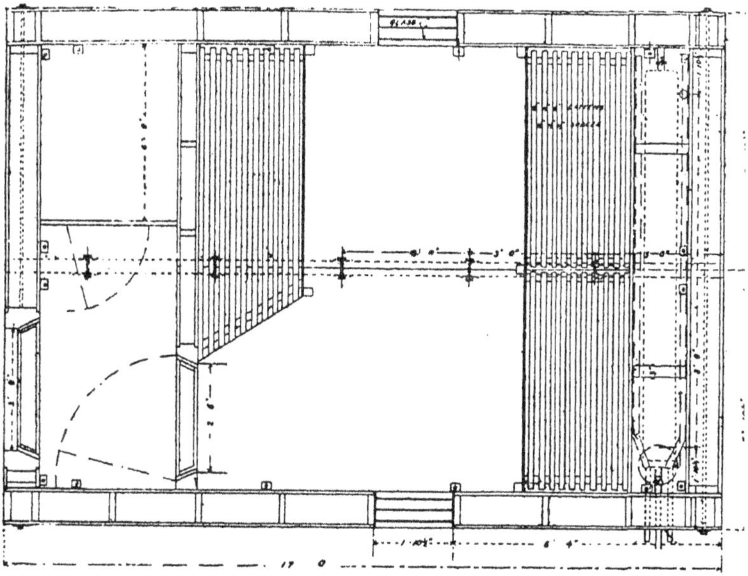

FIG. 106.

forming an air lock, a most important point if the entrance is at the side. The cooling is done by a brine-wall or screen placed at the back end, and through which the refrigerated brine circulates. The air in the chamber is cooled by contact with the sides of the brine-screen. When cooled it descends, and is replaced

[1] Messrs J. & E. Hall, Ltd., Dartford.

by warm air drawn from the top of the room, this being guided by the screen placed 3 inches from the ceiling.

In chambers of this size it is necessary to have a brine-wall or reservoir of this nature, so that the refrigerating effect may continue after the machine has stopped running. The cold chambers can be kept at any temperature either above or below freezing, and they are exceedingly useful for preserving game, fruit, etc., for use out of season, or for cold storing bulbs and root plants to retard their time of flowering.

# CHAPTER XIII.

## ARTICLES IN COLD STORAGE.

**Eggs.**—Eggs are exceedingly sensitive and readily absorb odours and smells. They are always giving off gas, and this must be removed as soon as it is formed or fungus will form and the egg will become musty. The egg storage room must be constructed of specially odourless timber, and is best cooled on the combination system by brine-pipes and air. It is essential that a gentle circulation and ventilation should be kept up to remove the gas and moisture given off by the eggs, and to prevent the formation of white fungus or mildew. The rooms should be, preferably, limewhited, and eggs should never be stored with other produce.

For storage purposes the eggs should be collected at the latter end of March and in April and May when they are finest, most plentiful, and in the most healthy condition. Eggs gathered in the summer are liable to be affected with heat-spots; these are small light-coloured spots which form on the yolk, and though an egg affected in this way need not necessarily be bad, it is unsuitable for cold storage and should be rejected.

Before being packed the eggs should be carefully sorted, graded, and candled. They should be again candled before they leave the store for the market. The packing-cases are best made of some white odourless wood, such as spruce or poplar, each case containing thirty-six dozen. The eggs are packed in fillers or light divisions, which fit into the cases and make a separate cell for each egg. These are made either of cardboard or wood pulp; the former is apt to smell if it gets damp and give the eggs a bad flavour, and the latter, though it gives the best results, is apt to warp and damage the eggs. The eggs should be placed in the filler resting on the small ends, as in this position the yolk will keep central better than in any other. If the egg is placed on its side the yolk is liable to shift and attach itself to the shell, making what is known as a "spot" egg. Both before the fillers are put in position in the cases and before the box is nailed down, a layer of coarse granulated cork should be put in to act as a cushion between the eggs and the covers. A material known in America as "Excelsior" is much used for fillers. It is made out of bass-wood shavings, and should be kept in stock some time before use in order that it may get seasoned and lose its odour.

Eggs for cold storage purposes should never be washed, as this removes a fine coating resembling varnish which naturally covers the shell and renders it less porous. If this varnish is removed, air gets to the meat and the process of decay becomes rapid. Unfortunately it is very difficult to tell a washed from an unwashed egg. The cases containing the eggs should be stacked in tiers with laths between them; this makes an air space round each case and allows efficient ventilation to take place. With the same object the cases are sometimes made with the sides an inch narrower than the ends.

The best temperature for eggs is 32° F., though for short periods of two to three months it may go as high as from 38 to 40° without damage. Some authorities hold that a temperature of 40° all the year round gives satisfactory results, but the consensus of responsible opinion inclines to the lower temperature. If the temperature falls below 28°, the yolk becomes hard and the egg is apt to freeze. At 30–32° the egg is in its normal condition with the white slightly thickened, and this helps to keep the yolk central. There is also smaller shrinkage and a lessened tendency to mildew and must. A temperature of 40° is cheaper to maintain, but at this the white is liable to become thin and the eggs are more liable to "spot" and become rotten.

One of the most important matters in the successful storage of eggs is the retention of the atmosphere at the proper degree of humidity. If the air is too humid the eggs will go mouldy; if too dry they will give up their natural moisture and shrink as much as 20 per cent. At 32° a humidity of about 70 per cent. is about right. Mr Maddison Cooper, who has devoted considerable attention to the subject, gives the following list of the proper degree of humidity for egg-rooms:—

| *Temp. in degs. F.* | *Relative humidity per cent.* |
|---|---|
| 28 | 80 |
| 29 | 78 |
| 30 | 76 |
| 31 | 74 |
| 32 | 71 |
| 33 | 69 |
| 34 | 67 |
| 35 | 65 |
| 36 | 62 |
| 37 | 60 |
| 38 | 58 |
| 39 | 56 |
| 40 | 53 |

The humidity can be easily ascertained by a wet- and

dry-bulb thermometer. When taken out of storage eggs should be tempered—that is to say, they should be removed from the egg-room at 32° to another room at 40° for, say, twenty-four hours; they should then be eased up to whatever the outside temperature is. This will prevent moisture depositing upon them and doing them harm.

There are several egg-testing machines on the market, the principle of which consists of rapidly passing the eggs over a strong light. Space, however, will not permit of a detailed description of them.

A handy candler is illustrated in fig. 107.[1] The egg is

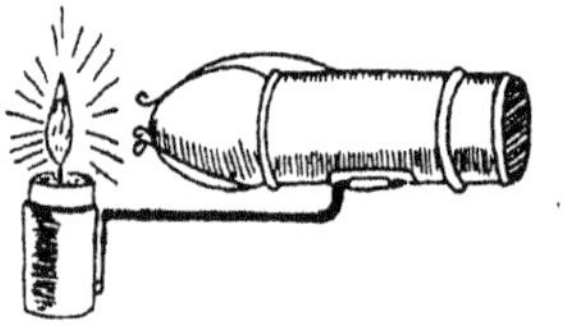

FIG. 107.

held in position by the spring clip and the tube is fitted with a lens. This is, of course, only useful for examination work; it would be too slow for commercial use. Candling is a fairly good test of purity, but it is not conclusive, and an egg may pass candling and yet be only fit for an election. Rotten eggs are largely used in tanning fine skins, and can often be disposed of for this purpose.

**Hops.**—Proper temperature, 26–28° F.

Hops when first harvested are dried in kilns, packed and sent to market. During the drying process some of the essential oils are driven off, and, if subsequently stored at ordinary temperatures, this process goes slowly on until the hops lose the peculiar aroma and flavour which

[1] *Ice and Refrigeration*, Chicago.

constitutes their value and which is given them by these oils. The drying process is essential, otherwise the hops would become crusted and useless; but it has been found that keeping them in cold storage retards the subsequent volatilization of the oil and retains their quality. The proper temperature for storage is from 26 to 28° F., and the tendency of the day is for a lower temperature still to be kept. The storage rooms should be constructed in the ordinary manner, and cooling may be done by either of the three systems, air blast being most common. There is no occasion for the rooms to be ventilated, and the same air can be circulated over and over again. The humidity should be kept at about 75, and the rooms should be perfectly dark. Great care must be taken in keeping the rooms absolutely dry, as the hops will be ruined if allowed to get moist.

Hops must also be carefully treated when brought out of cold storage to prevent moisture depositing upon them. There is no need, systematically, to defrost them by process; all they require is to be gradually tempered up to the temperature of the outside air.

Needless to say, cold air that has been in contact with hops must not be allowed to come in contact with other goods.

*Meat.*

**Frozen.**—Proper temperature, 22° F.

Meat when frozen is one of the simplest things to cold store. It is one of the few animal substances which does not gas, and it will keep for long periods without the air being changed. As a general rule the carcasses are packed one on top of the other just as they come in to the store, and their irregular shape allows the air to circulate round them and reach every portion.

Meat will *keep* for an indefinite period, but if stored too long becomes stale, goes off colour, and deteriorates in value as a marketable commodity. It is also desirable to subject it to changes of temperature as little as possible, and it has been found that meat killed and then stored on the other side before shipment does not keep its freshness so well as meat shipped directly after freezing. The shorter the period of storage the better, and any storage that must be done should be done near the point of consumption, not at the point of origin.

**Chilled.**—Proper temperature, 29 to 30° F.

Chilled meat will not keep for long periods—twenty-eight days being about the limit. This time is sufficient to enable it to be brought from the River Plate to Liverpool and sold, but up to the present it has not been found possible to bring it from Australasia. Chilled meat should in all cases be hung like fresh meat, and a variation of a degree one way or the other causes deterioration. For this reason brine circulation should be employed, and the rooms should be heavily piped so as to ensure an even temperature.

**Butter.**—Proper temperature, 16 to 20° F.

Butter is commonly put into cold storage in May and June, and is held until the following autumn, when it is taken out and put on the markets. In this country shipments are constantly arriving from Denmark, Holland, Canada, and other parts all the year round, the New Zealand season being in full swing when the Canadian is quiescent. There is thus a constant demand for cold storage, though, owing to the regularity of supply, the period of storage is not usually for long. Butter should be frozen hard, and is best stored in tubs just as it arrives, these being packed one on top of the other as they fit.

Butter is exceedingly sensitive to foreign odours and should be stored in a room by itself, care being taken that the room is absolutely sweet before it is put in. In the United States butter is sometimes held at zero, but it is becoming recognized that so low a temperature is not necessary. Occasional ventilation is desirable.

**Cheese.**—Proper temperature, 32 to 33° F.

Cheese follows much the same rules as butter. It is sensitive to odours, and should be stored in a room by itself. Ventilation is necessary. It should not be frozen or it will become "short" and crumble away. Cold storage has been found greatly to assist cheese in maturing, and large depôts have been erected by the Government in Canada solely for the purpose of ripening the cheese before it is shipped to England, the enhanced price obtained for it amply repaying the cost.

**Fish.**—The demand for fish storage is becoming greater and greater yearly. It is perhaps not commonly realized that the "fresh" fish consumed at the London dinner-table has probably been dead for a fortnight before it is cooked. All fishing-fleets carry their supply of ice, and the provision of this ice in a crushed state is the most profitable source of income to the ice factories at the principal fishing-ports. Fish preserved in this way, however, comes more under the heading of chilled than frozen, as the temperature never goes below 32°. Freezing fish is a difficult business, and fish are never frozen unless they are intended to be kept for some length of time. There are two ways of freezing. In one the fish are packed together in a tin case with water and frozen solid into a lump of ice; in the other they are taken into the freezing-room and, when frozen, are dipped one by one into water, which freezes on them and surrounds them with a thin film of ice. This ice forms a mechanical protection and allows them to be handled to any extent. Frozen fish can usually be detected from fresh fish, as the eye will dry up from the exposure to the cold, and the skin is liable to become hard and loose on the flesh.

Fish should not be stored with other produce. Brine or direct expansion cooling is most suitable.

**Oysters.**—Proper temperature, 35° F.

Oysters may be kept in cold storage for some time. They should never be frozen, or the shell will open. The cheaper grades are frequently taken out of the shell, put in buckets, and frozen solid. This, however, to a large extent, destroys their flavour.

*Vegetables.*

**Potatoes.**—Proper temperature, 40° F.

Potatoes are usually taken into storage in the autumn and preserved for spring and summer use. New potatoes are sometimes kept in this way. Cold storage prevents the eyes sprouting as the spring advances. Potatoes should be packed in loosely woven bags or in crates and stacked one on top of the other, so that the air can get all round them. If stored in bags the weight of the top bags should not come on the lower ones, but a rack should be provided. The air should not be too dry or shrinkage will take place.

**Onions.**—Proper temperature, 32–34° F.

If possible avoid storing onions, especially the coarser kinds. The better sorts can be preserved for six or eight months. They should be stacked like potatoes. Onion-rooms should be heavily varnished to prevent the odour getting into the timber. After storage, the room can be rendered pure by washing it down with a solution of calcium chloride. Brine circulation with occasional ventilation is the best.

**Cabbages.**—Proper temperature, 40° F.

Many kinds of vegetables store well and should be treated in a similar manner to the foregoing. Other vegetables can be stored for shorter periods. It would take too much space to go into the details of each. The temperatures recommended will be found elsewhere.

*Fruit.*

The Saccharine matter in fruit is of the nature of starch when the fruit is green, but as it ripens this turns to sugar. As soon as the fruit is quite ripe the sugar starts to ferment and the fruit begins to go bad. It is well known that fermentation is promoted by heat and retarded by cold, and the natural deduction is that a cold temperature will preserve fruit.

With the object of ascertaining the proper temperature and best conditions for the storage of fruit or vegetables, a series of experiments was carried out by the Kansas State Experimental Station in 1900. The following is a list of recommendations :—

| Product. | Temperature. | Best packages. | Period. |
|---|---|---|---|
| Apples (summer), . . | 38–42 | Barrels or boxes. | 2–4 months |
| ,, (winter), . . | 32–35 | ,, ,, | 5–8 ,, |
| Pears, . . . . | 33–38 | ,, ,, | 2–4 ,, |
| Peaches, . . . | 36–38 | Crates. | 2–4 weeks |
| Grapes, . . . | 38–40 | Sawdust boxes. | 6–8 ,, |
| Plums, . . . . | 38–40 | Crates. | 2–4 ,, |
| Cherries, . . . | 40 | Ornate boxes. | 1–3 ,, |
| Water melons, . . | 40 | ,, ,, | 3–6 ,, |
| Musk melons, . . | 40 | ,, ,, | 2–3 ,, |
| Tomatoes, . . . | 38–42 | Crates. | 2–4 ,, |
| Cucumbers, . . . | 38–40 | ,, | 1–3 ,, |
| Cranberries, . . . | 34–38 | Barrels. | ... |
| Onions, . . . | 34–40 | ,, | ... |
| Potatoes, . . . | 36–40 | ,, | ... |
| Cabbages, . . . | 34–35 | Crates or barrels. | ... |

With a similar object the Technical Educational Committee[1] of the Kent County Council carried out in 1898 a series of experiments at Messrs J. & E. Hall's works at Dartford. Mr W. P. Wright, County Superintendent of Horticulture, conducted the experiments, and his report has been reproduced in most of the trade journals. Generally speaking, the results obtained agreed with the foregoing, and it was found that fruit kept best at temperatures of from 32–36°, which is rather lower than those recommended by the Kansas station. The bulk of the fruit was put into storage in September and taken out between that date and the end of the year.

In the majority of cases the fruit was covered by a layer of waterproof paper or cotton wool, and with the softer fruits this was found to be beneficial.

In neither of the reports is it stated what precautions were taken to ensure regular ventilation, a point of the greatest importance.

**Apples.**—The summer or soft apples are more perishable and require a higher temperature than the winter ones. They can only be stored successfully for a comparatively short time. Firm winter apples can be kept for as long as two years, though eight months is about the commercial limit. Experiments conducted by the Nebraska State Horticultural Society showed that apples keep best if each fruit is wrapped first in a sheet of waxed paper and then in a sheet of newspaper, finally being packed in barrels and stored at a temperature of 36°. Treated in this way, the fruit kept for over a year in first-class condition and gave much better results than similar fruit put naked into barrels.

[1] Kent County Council, 29th Report of the Technical Education Committee.

On the other hand, the Dartford experiments showed that no covering was necessary; the storage in this case, however, was not of long duration.

For successful storing the apples should be hand-picked off the trees, carefully sorted and graded; they should be carefully packed into their barrels and kept "this side up" if possible.

At Dartford the early dessert apples—the summer apples of the Kansas experiment—which gave the best results were as follows:—Blenheim Orange, Chelmsford Wonder, Colonel Vaughan, Cox's Orange Pippin, Duchess's Favorite (*alias* Duchess of Gloucester or Scarlet Incomparable), and Worcester Pearmain. Most of these were in good condition at the end of four months.

**Pears.**—For storage, pears should be picked under-ripe, wrapped in tissue paper, and carefully packed. As in the case of apples, the Dartford experiments found coverings of no value, but this does not agree with other results obtained.

The softer varieties should be stored at from 36–38°, the harder kinds 3° lower. It was found at Dartford that Beurré D'Amanlis, Conference, Doyenné du Comice, Fertility, Pitmaston Duchess, Williams Bon Chrétien (the American Bartlett) kept well for some four months. Beurré Hardy would probably have kept, but the fruit arrived in a bruised condition. Williams pears, as is well known, rapidly become sleepy, and keep best when picked under-ripe.

**Soft Fruits.**—All the softer juicy fruits will not keep for any length of time in cold storage. The only object in storing them at all is to keep them during transport or to hold them over a glut in the market for a week or a fortnight. To obtain good results with fruit of any sort, but especially with these, it must be slightly under-ripe and perfectly sound when stored, the slightest bruise being fatal. This necessitates great care in picking, sorting, and packing; but, as a rule, the prices at the end justify the trouble.

Strawberries, red and black currants, and cherries, when tried at Dartford, kept in good condition for from four to six weeks.

Fruit should never be put direct from a summer temperature into cold storage, but should be gradually tempered down; the same process should be gone through when it is taken out. If cold stored fruit is suddenly exposed to a hot atmosphere, "sweating" will occur and it will be ruined.

**Peaches.**—Except the harder clingstone varieties, these are not well adapted for cold storage. The best results have been obtained by wrapping the fruit in waterproof paper and packing it in well-ventilated egg-boxes. Dartford experiments found that a temperature of 32° F. was best, and that Royal George, Crimson Galande, and Lady Palmerston kept well for some two months, but the flavour was indifferent.

**Grapes.**—These must be absolutely dry. The best results have been obtained by packing them in sawdust, but hanging the bunches in a dry atmosphere also answers well for some six or eight weeks.

If the air is too dry shrinkage will occur. The fruit must be carefully watched, and the humidity regulated by results. In the Dartford experiments it was found that Black Hamburgh and Muscats kept at a temperature of 32° for some two months; the bunches wrapped in waterproof paper came out best.

**Plums.**—The more juicy kinds and cooking varieties go

off rapidly, but the firmer varieties will keep for some four to six weeks. At Dartford it was found that a temperature of 32–36° was best, and good results were obtained from Greengages, which kept for ten weeks, and Orleans, Diamond, and Sultan, which kept for from four to six weeks. Victoria kept nine weeks. Monarchs, which were too ripe when stored, only kept a month; stored under-ripe they would probably do well. Coe's Golden Drop kept two months and retained its delicious flavour though the skin was slightly shrivelled.

**Tomatoes.**—If picked just as they are beginning to redden and when still hard, wrapped in tissue paper, and carefully packed in a crate, tomatoes will keep for about two months. If picked green and stored, they will not ripen when taken out of cold storage. The fruit should have the foot stalks removed before being put into storage, as these are apt to start decay.

**Oranges and Lemons.**—Proper temperature, 45–50° F.

Both these fruits should be put into storage while still unripe. They give off great quantities of gas, and a steady ventilation should be kept up in the storage rooms. The air must not be too dry or shrinkage will occur. Oranges can be frozen solid, and in this state they are considered a delicacy in very hot weather. Once frozen they go bad very rapidly if thawed out.

**Nuts and Dried Fruit.**—Proper temperature, 40° F.

These are principally stored during the hot weather to preserve their flavour in the first instance and to preserve them from the attacks of grubs in the second. Dried fruit, such as raisins, figs, prunes, and dates, should go into storage in May and be retained there until October.

**Tobacco and Cigars.**—Proper temperature, 40–45° F.

Fine leaf tobacco is very sensitive, and, like hops, loses its flavour and aroma if exposed to the ordinary atmosphere for any length of time. The object of storing it is to retain this, and at the same time to guard it from the attacks of insects. Tobacco of all sorts is rapidly ruined if it gets alternately moist and dry, as it will do with the altering humidity of the ordinary atmosphere.

**Tea.**—Though, as far as is known, tea has not yet been stored in any great quantity, there is good reason to suppose that the flavour would be retained better, especially in the finer grades, if stored at an even temperature than if exposed to the variations of the ordinary atmosphere. That tea is exceedingly sensitive is shown by the fact that tea brought overland is superior in flavour to that brought by sea, even when this latter is packed in lead-lined cases.

At a certain stage in the manufacture of tea—*i.e.*, when it is being fermented—refrigeration is very desirable, as in the hot moist climate of the tea-producing countries it is very difficult to control the temperature with water alone. Also the excessive humidity of the atmosphere prevents the leaf shrivelling properly.

**Furs and Woollen Goods.**—Proper temperature, 35° F.

The object here is, of course, to prevent the attacks of moths. There are several fur stores now in existence, and several large woollen houses in the United States have had their warehouses fitted up with refrigerating machinery in order to preserve their winter woollen goods from the attacks of vermin during the summer months. It is probable that this practice will become more general. In Russia it is the universal practice to place fur coats in refrigerated houses during the summer, but in this case the cold store usually carries on a pawnbroker's business as well, and advances money against the value of the coats stored. British cold storage houses have not yet developed this branch of the business to any great extent.

# CHAPTER XIV.

## OTHER APPLICATIONS.

### *Refrigeration in the Brewery.*

**The Objects of Refrigeration.**—The main object of placing a refrigerating machine in a brewery is (1) rapidly to cool the wort after it comes from the coppers and before it is pitched with yeast for fermenting, and (2) to cool liquor for attemperating the wort during fermentation. There are, of course, other purposes for which refrigeration is now regarded as essential, especially in lager beer breweries, such as for cooling yeast-rooms, cellars, hop stores, etc.; but the two purposes mentioned above are the most important. A refrigerating machine as used in a brewery is practically the same as that used for any other purpose; the only difference is in the application, and before this can be studied it is necessary to know something of the nature of the wort to be cooled, and also of what goes on in the vats during fermentation, or, as it is sometimes called, attenuation.

**The Cooling of Wort.**—Wort is the liquid resulting from the mashing of malt and the boiling of the extract with hops. It is rich in nitrogenous matter, and therefore readily becomes acid or putrid if allowed to stand after it has been boiled. For this reason it should, as it comes from the hopback, be at once cooled and impregnated with pitching yeast of vigorous growth, specially cultivated for the purpose. It is desirable to cool wort in contact with air, as at high temperatures it will take oxygen into combination, and by that means throw off matter held in solution; and it is in consequence of this property that aerated wort will eventually become bright and clear, a result unattainable when air has been excluded during cooling. Some brewers pass the wort, as it comes from the hopbacks, directly over the refrigerator, first giving it a slight fall in order to aerate it; this method, however, requires large refrigerating power, and the more usual course is to pump it first over the cooling floor—which consists of shallow vessels, sometimes of wood, but preferably of flush-riveted copper—where it loses some of its heat by radiation, and so does not make such a large demand on the refrigerating machine.

At high temperatures wort is very susceptible to the attacks of lactic acid or base ferments, and, unless the temperature is rapidly reduced, these multiply and ruin the beer. In order, further, to guard against the base ferments, the coolers should be kept thoroughly clean, and the joints of the woodwork, and the pipes, should be free from dirt and albuminous deposit, which would foster their growth rather than that of alcoholic yeast.

The refrigerator is supplied with refrigerated or iced liquor cooled to a temperature of from 40 to 50° F., or, where liquor is scarce and has to be bought from the water companies, to a temperature as low as 33° F. In the rapid cooling over these refrigerators the wort deposits the matter previously held in solution, and this matter is left adhering to the cooler. The temperature of the wort as it leaves the cooler is, with pale ale, from 55 to 60° F., and at this temperature, or at whatever is normal to the system in use, it runs off to the fermenting vats, where, as before stated, it is at once pitched with yeast.

**The Fermentation of Wort.**—In fermentation the action which takes place is much more complex; in fact—though that is hardly a question to be gone into here—it was not understood by brewers themselves until Pasteur developed and proved the theory that fermentation was due to the development of the microbes or cell life of which the deposit in wort and beer largely consist. Before pointing out the action of fermentation, however, it will be necessary to go back slightly and enquire what causes it. During the conversion of barley into malt, a substance is formed which is known as "diastase"; this converts the insoluble starch of the barley into a soluble and fermentable sugar, and this sugar, when absorbed as oxygen by the yeast, splits up into nearly equal proportions of $CO_2$ and alcohol. The true yeast microbe is a special form of cell, whose normal condition of existence is one removed from the air—*i.e.*, anaërobic; it feeds on the nitrogenous matter present in the wort, and breathes the combined oxygen of the sugar of the malt, splitting it up as described. The *fermentable body*, therefore, is the one capable of giving up its oxygen, and the *ferment* is the microbe having the power of taking it. Under the action of this microbe the excess of nitrogenous matter is absorbed, fermentability is decreased by the transformation of the sugar, and the beer is made stable by the addition of the alcohol in an active form that was previously present in the maltose in a passive one.

This action of fermentation is accompanied by heat, and the rise in temperature varies with the character of the beer and the system of fermentation in use, the ordinary rule for English breweries being a rise of about 1° F. per pound of gravity. The temperature at which it is necessary to keep the wort is easily found when it is remembered that it is not considered safe or desirable to exceed, at any rate for pale ale, the limit of 70° F. The brewer with this class of ale, therefore, would start at a temperature of, say, 58 or 59° F., and would gradually rise, attaining a temperature of 68 or 69° F. by the time fermentation was complete. With strong beers containing a large quantity of malt, and in which the rise per pound is higher, fermentation is started at a lower temperature, in order that the final temperature may not exceed the limit of 70° F. Of course, with very strong beers containing an excess of alcohol this limit of 70° F. may be exceeded without danger, as, owing to the presence of the alcohol, its stability is greater. After fermentation the beer is usually held at the same temperature for about five hours; it is then gradually thrown back at the rate, at first, of a degree in five hours, then another degree in, say, four hours, the rate getting quicker and quicker until the racking temperature is reached.

It may be remarked in passing that the speed of fermentation, which, however, is more a matter for the brewer than the engineer, has a distinct bearing upon the character of the beer. High temperatures are said to give a peculiar flavour which is much sought after, but rapid attenuation—and, consequently, a high temperature—is liable to cause difficulties in clarification and a lowness of stability owing to the formation of weak alcohols.

**Heat to be Removed.**—The exact quantity of heat to be removed from the wort during cooling varies with its sensible and specific heat and its specific gravity. Ale wort, as brewed in England, is not cooled below a temperature of from about 60 to 55° F., and the storage rooms are not kept much below the same temperature. With lager and other weak beers, on the other hand, much lower temperatures are necessary.

The heat to be removed during fermentation varies with the specific gravity of the wort, or, in other words, with the amount of malt or saccharine matter that has been used in its manufacture, and much more refrigeration is required for strong beers containing much malt than for weak beers containing little. These quantities are easily calculated from well-established formulæ for various strengths of beer, and the size of the refrigerating machine installed depends on the result. The wort, however, has to be cooled very rapidly, and, therefore, a large quantity of heat has to be removed in a very short time; this is the most important factor in determining the size of the machine. The refrigerating capacity required for cooling yeast-rooms, cellars, etc., is ascertainable on ordinary cold storage rules, and presents no special features or difficulties; but it must be remembered, in this connexion, that fermentation is not usually completed in the fermenting vats, and that the beer still continues to give off heat when it is standing in casks in the cellars.

The specific heat of wort decreases as the specific gravity rises. This latter is usually ascertained among English brewers by the "pounds per barrel" saccharometer, and as specific gravity shows how much heavier or lighter a liquid is than water which is taken as 1000, so saccharometer pounds show the weight of a barrel of wort in excess of that of a barrel of water which weighs 360 lbs. There are several saccharometers in use, each having its own table of constants; as a rule, they are correct at 60° F., and a constant figure has to be added for every degree the wort is in excess of that temperature. The average brew of ale wort will show 19·8 on the pounds per barrel saccharometer corresponding to a specific gravity of 1055, which is the excise standard, and to a specific heat of about ·905; during fermentation it will lose about 40, and when ready for racking will show about 1015 (5·4 pounds per barrel). Stout will probably show 1090 (32·4 pounds per barrel), and will be brought down to 1035 (12·6 pounds per barrel).

**During the Cooling of the Wort.**—The wort loses a certain amount of heat by radiation in the hopback and the pipes through which it passes to the cooling floor, so that by the time it reaches the refrigerator its temperature is about 140° F. Sometimes, however, it is passed from the hopback directly on to the refrigerators; in this case it reaches them at a temperature of about 190° F., and, if pale ale, is reduced by them to a temperature of from 55 to 60° F. The general arrangement of the refrigerators will be described more fully subsequently. For the present it is sufficient to say that they are generally made in two portions. The upper coils are supplied with well or water companies' liquor or water, which forecools the wort and reduces its temperature to about 70° F., and the lower coils with iced liquor, refrigerated brine, or (more rarely in England) by directly expanded ammonia, which further reduces the temperature. The term iced liquor is a survival of the days prior to the introduction of mechanical refrigeration, when ice used to be bought and put into the cold liquor back or tank to reduce the temperature of the cooling liquor. The term has survived, though the methods have changed. Where liquor is scarce it is sometimes reduced to a temperature of from 33 to 34° F., and fed through all the coils of the refrigerator, entering at the bottom and leaving at the top. In any case the total heat to be removed from the wort is the same, and, having ascertained the average strength of the wort brewed, and, therefore, its specific weight or gravity and heat, the number of British thermal units to be removed would be—

$$\text{B.T.U.} = B \times 360 \times g \times s\ (t - 40),$$ [1]

in which $B$ = number of barrels to be cooled; 360 = the number of pounds in a barrel ($36 \times 10$); $g$ = the specific gravity of the wort; $s$ = the corresponding specific heat; and $t$ = the temperature to which the wort is cooled by the upper coils which are fed with well or water companies' (unrefrigerated) liquor.

If it is required to reduce the British thermal units to tons refrigeration, the product of the above formula must be divided by 318,080,[2] being the number of B.T.U. absorbed by 1 ton of ice melting from 32 to 32° F. For preliminary estimates, however, it is generally sufficient to take the number of lbs. of wort to be cooled (barrels × gallons × lbs.), multiply them by the degrees cooled mechanically, and divide the product by 318,000. If all the liquor used in the refrigerator is iced, the former formula must be modified accordingly.

The essence of the successful cooling of wort is rapidity, and, therefore, the actual size of the machine capable of doing this must be considerably greater than the above results would suggest. The rate at which a given quantity of wort can be cooled is obviously governed by the number and size of the refrigerators available, each refrigerator

[1] Seibel; adapted for British units.

[2] For American practice this figure would be 284,000.

only being capable of efficiently passing a certain quantity of wort in a given time. To find the actual size of the machine required for wort-cooling, therefore, the tonnage given above must be multiplied by 24 divided by the time in hours which the refrigerator takes to pass the wort.

**During the Fermentation of the Wort.**—After fermentation there is a slight decrease in the quantity of the wort, due to the consumption of the yeast-forming matter; it is not great, however, and hardly affects the total. The heat generated by 1 lb. of maltose while splitting up under the action of the yeast into alcohol and $CO_2$ is approximately 330 B.T.U. In order to find accurately the heat in B.T.U., $H$ to be removed by the attemperators during fermentation, the specific gravities of the wort ($b$) and of the beer ($b_1$) must be found. The following formula can then be used—

$$H = \frac{B \times 0{\cdot}91\,(b - b_1)\,(360 \times b)\,330}{100},$$ [1]

or, if in tons refrigeration, the product must be divided as before by 318,080.[2]

As, however, the specific heat of wort is so near that of water, it is sufficient for preliminary estimates to let

$$H = B \times 360 \times 330 \text{ B.T.U.}$$

As the amount of saccharine matter in the beer becomes less, the heat to be withdrawn during fermentation decreases, and the smaller will be the refrigerators required.

**Other Duties of the Refrigerating Machine.**—The remain-

[1] Seibel; adapted for British units.

[2] For American practice this figure would be 284,000.

ing duties to be performed by the refrigerating machine are the cooling of the cellars, yeast-rooms, hop stores, etc. It is quite impossible to lay down any formulæ for this work, as the refrigeration required depends on so many factors, which in practice have a constantly varying value. The most important factor is, of course, the insulation of the walls of the cellars and rooms, but the frequency with which they are entered and the temperature of the beer stored have also to be considered. Cellars are usually kept cool by piping as opposed to air blast, and the amount of piping required depends on the above factors, and, in addition, on the initial temperature of the medium circulated through them. It is, therefore, absolutely necessary in dealing with these particular items to take each case on its own merits, and to lay down no general rules, which will only mislead and cause loss and trouble.

For instance, American writers have stated that as a rough figure the refrigeration tonnage required in a brewery may be obtained by dividing the number of barrels brewed per day by 4, and that in cooling wort 1 ton of refrigeration is required for 40 barrels, while for removing the heat of fermentation 1 ton is required for 35 barrels. These are not set out as accurate or reliable rules, but the danger of accepting them, even approximately, for British practice will be realized when it is remembered that in the United States a barrel is 31 gallons of 8·3 lbs., and a ton refrigeration represents 284,000 B.T.U.

**Types of Wort-Coolers.**—The various types of wort refrigerators, both vertical and horizontal, used in breweries are well known and need not be described. The Baudelot cooler, pure and simple, has now largely gone out of use for this purpose, its place being taken by refrigerators with flattened pipes, giving a better interchange of heat.

The horizontal cooler is not often used alone, but frequently in combination with a vertical one. In this case it is usual to arrange them in the form of the letter L. The hot wort flows down the vertical part of the cooler, where it is aerated, and then over the corrugations of the horizontal portion. The cooling liquor pursues the opposite course, entering the horizontal portion at the bottom, and—if the temperature range is great enough to stand it—leaving the vertical portion at the top. The warm liquor thus meets the hot wort. Usually, however, the vertical portion is supplied entirely with well water at a temperature of about 60° F. or whatever it may be, and the horizontal portion with iced liquor at a temperature of about 40° F. or lower, as the case may be. Where only vertical refrigerators are used, as is the case in perhaps the majority of breweries, either the whole refrigerator is supplied with iced liquor, or the upper tubes are supplied with well liquor and the lower ten tubes or thereabouts with iced liquor.

**The Direct Expansion Wort-Cooler.**—In America and on the Continent, where weak beers are the rule rather than the exception, lower temperatures all round are carried. The storage cellars are sometimes kept at 32° F. and the fermenting rooms at 36° F. It is now common practice, in view of these low temperatures, to use well liquor in the upper coils of the refrigerator, and to cool the lower coils either by the circulation of cold brine or by expanding ammonia directly through them as shown in fig. 108. The average English brewer would, and does, consider this dangerous, but it is really not so, as there is practically no chance of an escape of brine or of ammonia gas into the wort. As ammonia attacks copper, this material cannot be used in the refrigerator, therefore the lower coils have to be of iron or steel. In order to conform to custom and to prevent rusting, the practice at first was to cover these steel pipes with a thin sheet of copper drawn on over the iron pipe. In some cases, however, it was found that, owing to imperfect work, an air space formed between the iron pipe and its copper covering, preventing a proper interchange of heat, and thus reducing the efficiency of the apparatus. As a result, and after experiment, the copper covering has been discarded altogether, the iron or steel pipes being ground bright, polished, and used direct. The action of the wort running over the cooler produces a skin on the iron which acts like a varnish, and, as it does not deteriorate, effectively prevents any tendency to rust. In fact, so far as this matter is concerned, there is no reason why the whole cooler should not be made of bright iron or steel tubes.

There is, however, a practical objection to cooling by direct expansion. Very few ale breweries require or have separate machines or compressors for wort-cooling only, the usual arrangement being the installation of one or two machines for all purposes. Now, as wort-cooling represents about three-fourths of the total refrigerating work required, and, moreover, has to be done in a very short time, it is necessary perhaps to turn the whole refrigerating capacity of the brewery on to it, leaving the cellars, etc. to look after themselves for the time. As the wort does not usually require to be cooled to a very low temperature, the temperature of the expanding ammonia in the lower coils need not be lower than, say, 30 to 35° F., corresponding to a gauge back pressure of from 45 to 51 pounds per square inch—the temperature depending on the strength of the ale brewed; and, as it is a direct gain to have the back pressure as high as possible, this is an economical advantage. The cellars, however, and more particularly the liquor-coolers, are probably being run at a lower temperature, and, therefore, the gas returning to the refrigerating machine from the expansion pipes in them is at a lower pressure than that coming from the wort-cooler. This introduces difficulties which are not easy to overcome, and if, as before stated, the whole power of the machine is put on to wort-cooling for the short time it is in progress, the other portions of the brewery requiring refrigeration are left without it, and the temperatures begin to rise.

It has as a result been found advisable, where this class of wort-cooler is employed, either to connect the front and back ends of the compressor separately, and work them at different back pressures with direct expansion everywhere, or to use brine circulation in the cellars and yeast-rooms. Though this latter method, by introducing an extra interchange of heat, is not directly so efficient as the direct expansion system, it is much more easy to manipulate. The cooled brine is stored in a tank, and is then pumped from it through the pipes in the cellars. A reserve of cold brine is thus produced, so that when the whole refrigerating capacity is put on to wort-cooling, the brine-pump can still circulate sufficient cold brine drawn from the tank to keep the rest of the work going. As the ammonia is expanded directly into the coils of the wort-coolers, the ammonia gas returning to the compressor is all of the same pressure; the machine is working at its highest efficiency, and no difficulty arises. The expansion valve or valves by the coolers will have, however, to be carefully regulated, especially when top feed is in use, or the liquid ammonia will be freezing back to the compressor, and giving trouble in that way. When brine is used, the temperature in any particular room or place can be varied at will by the use of valves which control its circulation through the pipes. Brine has also been used in this way for wort-cooling instead of the direct expansion of ammonia. In this case about 20 per cent. more cooling surface has to be allowed, but the whole of the cooler may be of copper pipe.

**Attemporators.**—The heat of fermentation is removed by the attemporator, which is a coil of tinned copper pipe, usually of a flattened section, placed in the fermenting vat. Through this coil of pipe, the length of which varies according to the strength of the beer, iced liquor is circulated, taking up the heat. Sometimes refrigerated brine is used for this purpose as well, but sweet iced liquor is almost invariably preferred, especially where the attemporators are of the swivel removable type, as being on the whole better for the purpose. This liquor is cooled in a submerged cooler, or by letting it flow over the coils of a Baudelot cooler in which ammonia is expanding or brine circulating; it is then pumped to a tank placed at a higher level than the fermenting vats, and allowed to flow by gravity through the coils of the attemporators, the exact amount passing, and so the temperature being regulated as desired, by valves placed against each vat. Sometimes the fermenting tuns are covered in and brine-coils are arranged inside the top of the covering.

**Submerged and Baudelot Coolers.**—Carbonic anhydride being a natural product of the brewery, it is only to be expected that brewers would, other things being equal, favour any machine using this gas as a refrigerating agent on account of the harmlessness of any escape which might occur; and, acting probably on this reasoning, a large number of $CO_2$ machines have been installed in breweries. So far as the duties to be performed by the refrigerating machine are concerned, it is a matter of indifference whether ammonia or $CO_2$ is employed as the refrigerating agent, and the methods used in distributing the refrigeration produced by a $CO_2$ machine do not differ materially from those already described. When cellars or hop stores are to be cooled, the expanding $CO_2$ cools brine, which is circulated in pipes, or, if air circulation is desired, the ordinary brine air battery is used.

With $CO_2$ refrigerating machinery liquor-coolers of both the submerged and Baudelot type are employed. The submerged cooler is, generally, of similar construction to the submerged condenser, and consists of an insulated tank, in which the evaporating coils are placed and surrounded by the liquor to be cooled. The manifolds of the coils protrude at the top and bottom of the cooler, and the ends of the coils pass through glands in the casing, so that none of the joints are submerged or out of sight. It is considered by some authorities that the submerged cooler is more efficient than the Baudelot type, because in the latter there is a loss due to the warm air which comes in contact with the cold pipes and cooled liquor being chilled, and thus uselessly absorbing some of the power of the machine. This, however, will not occur if the coolers are, as they should be, inclosed in an insulated chamber.

On the other hand, it is urged by the advocates of the Baudelot cooler that as an elementary law of the transmission of heat states that the rate of transmission of heat from a cold surface to water increases in a direct proportion to the rate at which the water passes over that surface, submerged coolers can hardly be as efficient as the Baudelot type, on account of the more sluggish circulation of the liquor to be cooled. They also urge the further objection that, if the temperature is at all low, ice is liable to form on the expansion coils, which is, of course, all against the machine. This latter eventuality may, and does, happen with the open type cooler; but, owing to the greater velocity with which the liquor passes over the coils, lower temperatures may be reached before the ice begins to form, and, when it does, it can be more readily detected and removed.

**The Distribution of Liquor over a Baudelot Cooler.**—In cooling liquor or wort in a cooler of the Baudelot type, the even distribution of the liquor over the cooler becomes a matter of supreme importance, as it affects directly both the efficiency of the machine and that of the cooling. Generally speaking, the same principles apply as those which have to be observed with open-air condensers. The wort or liquor must flow down the coils in a thin even film, and there must be no splashing or spattering. There is obviously a limit to the amount of liquor that will flow over the coolers without this happening, and the limit may be taken at about 100 gallons per lineal foot of distributing pipe per hour. In practice half this quantity, or 50 gallons per foot per hour, is all that is usually allowed for efficient work. The actual pipe surface employed for cooling depends, however, very much on the temperature to which the liquor has to be reduced. Much more surface has, for instance, to be employed if the liquor is being reduced 16° from, say, 50 to 34, than if it is being reduced the same number of degrees from 70 to 54.

In order to distribute the liquor over the coils, the practice, until recently, was to have a slotted pipe running the length of the cooler, and to introduce the liquor at one end of this, letting it run along and overflow through the

FIG. 108. *To face page 375.*

slot over each side of the coils. This method, however, led to a very uneven distribution, that part of the cooler nearest the supply getting too much liquor, and that near the far end too little. Further, when the liquor was supplied to the cooler, as it often is, by a pump, the water would surge along at each stroke of the pump and would flow down the coils in waves. To overcome these defects, the liquor, in more recent practice, is introduced into a small tank placed in the centre of the length of the cooler, and is allowed to flow by gravity into the slotted pipe, and so over the coils. This method has got rid of the pulsation trouble, and has secured a much more regular distribution; but, even with it, if the cooler is at all long, the centre gets too much and the ends too little. The difficulty may be best met by having two or more tanks placed at even distances along the length of the cooler, the supply being regulated so that each part of the cooler gets its proper amount of liquor.

**Iced Liquor, and Direct Expansion, Baudelot Cooler for Wort and Liquor.**—A Baudelot cooler for cooling wort is illustrated in figs. 108, 109, and 110.[1] The upper coils, which are of flattened copper pipe, are cooled by well or iced liquor, and the lower coils, which are of 2-inch bright steel pipe, by the direct expansion of ammonia. In accordance with the law that at a given temperature, range, and velocity, more heat is transmitted at the higher temperature, and in order that the wort, as it is cooled, may come in contact with still colder liquor and so maintain the temperature range, the cold liquor enters these coolers at the bottom and leaves them at the top. The liquid ammonia is fed in at the top of the expansion pipes, and, evaporating as it runs through the coils, leaves

[1] Messrs L. Sterne & Company, Ltd., Glasgow.

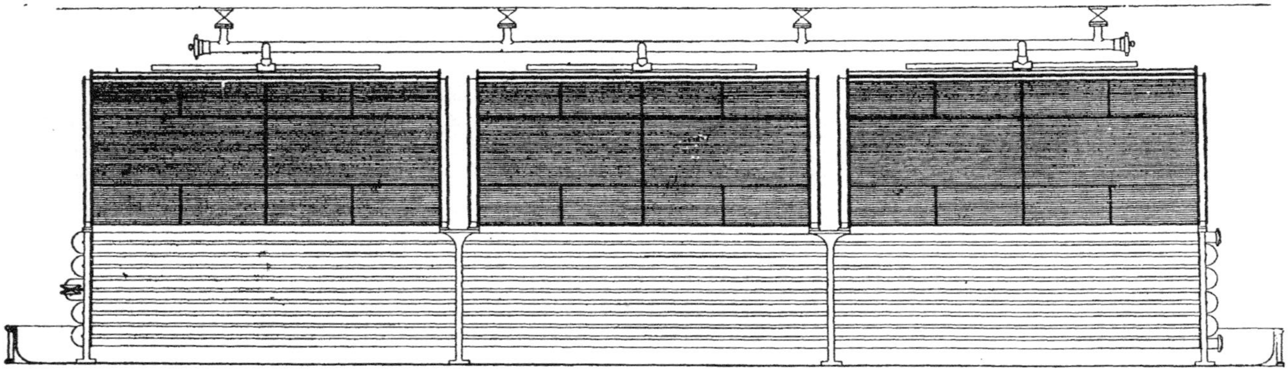

Fig. 109.

them at the bottom. This latter is known as top feed, and,

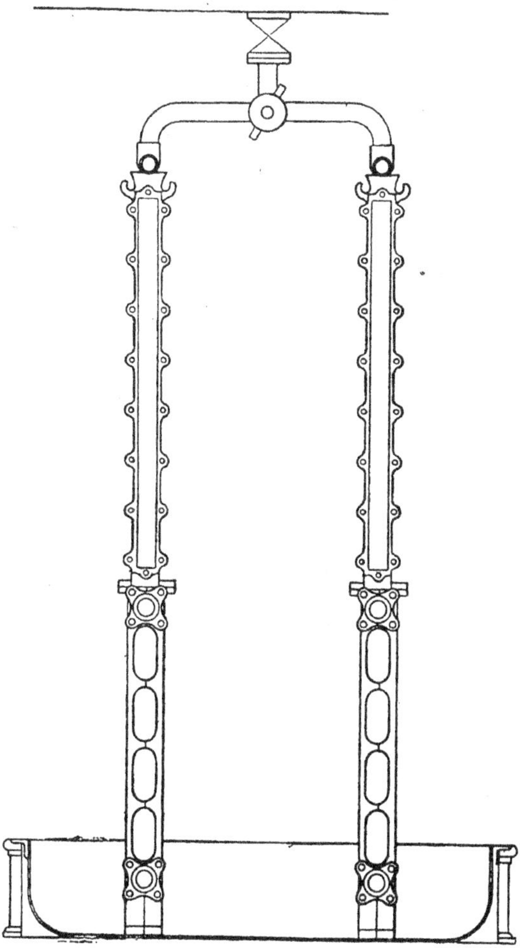

FIG. 110.

though in careless hands there is some chance of the ammonia reaching the compressor in an oversaturated state, causing it to act in exactly the same way as a steam cylinder does when the steam supplied to it contains water, the advantages of top feed are so obvious, and the risk is, comparatively speaking, so small, that it is nearly always adopted for this work. The same remarks apply to fig. 111,[1] which is a Baudelot cooler for cooling liquor, for wort-cooling, and attemperating.

The capacity of the two coolers illustrated is as follows:—That shown in fig. 109 is designed to cool 200 barrels of wort from 180 to 40° F. The well or iced liquor passing through the upper part of the cooler is at a temperature of 45° F., and reduces the wort to 60° F. The lower portion, with ammonia expansion, reduces it from 60 to 40° F. That shown in fig. 111 is for cooling liquor, for wort-cooling, and attemperating. The combined capacity of three coolers similar to that illustrated, is 100 barrels of liquor per hour cooled from 54 to 34° F. When erected, they are surrounded by an insulated casing in order to keep the warm air of the brewery from playing on the coils and giving up its heat to them.

If, instead of expanding ammonia directly into the lower coils, it is preferred to circulate refrigerated brine, the same construction as shown in the figures is adhered to, the cold brine being fed in at the bottom and leaving at the top of the liquor-cooler, and at the top of the ammonia-cooled section of the wort-cooler. Large numbers of this type of cooler using the latter medium have been installed in various breweries and have given good results.

**Chilling, Filtering, and Racking.**—The final duty to be performed by the refrigerating machine is the cooling of the ready beer before racking. The old-fashioned settling

[1] Messrs L. Sterne & Company, Ltd., Glasgow.

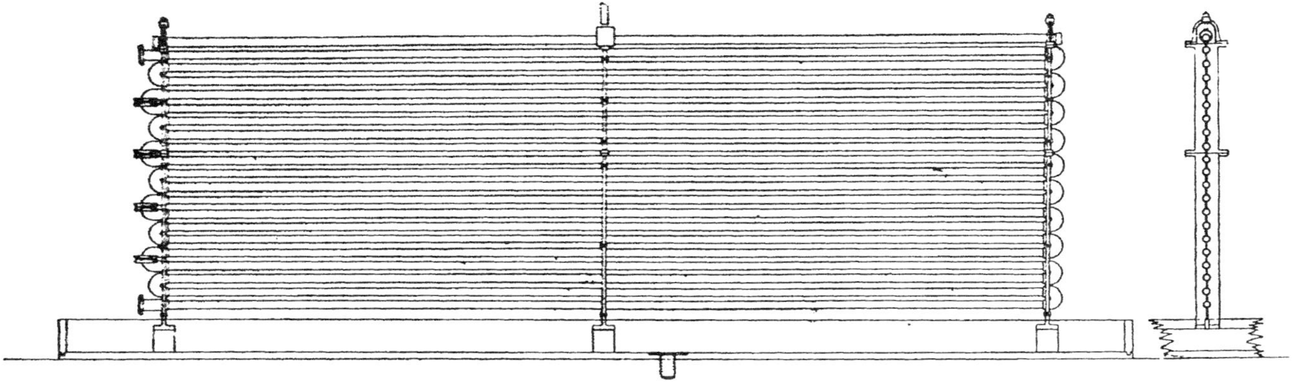

FIG. 111.

tank has now, in the majority of breweries, become the racking tank. When the beer leaves the fermenting tuns, it is run into a shallow skimming back, where excess yeast is skimmed off. In most breweries the beer is racked direct from the racking back, and allowed to stand in the casks in the cellars for some time before shiving. Sometimes, however, it is chilled before being racked. In this case the beer must, after chilling, be passed through filters so as to get rid of the albuminous deposit which is thrown down by chilling. If this is not done, the beer, on a rise of temperature, will reabsorb the deposit and become turbid and cloudy. The coolers used for this purpose vary; some consist of pipes enclosed in a drum through which refrigerated brine is passing in an opposite direction to the beer which runs through the pipes. This type of cooler has also been suggested for use in wort-cooling, but, as the wort would not be in contact with the air—a condition, as has been shown, essential to its clarification—it is exceedingly doubtful whether its use would be attended with any success.

**Lager Beer.**—The brilliancy and sparkle which is much admired in lager beer is due very largely to the manipulation of the temperature during fermenting and racking. In lager beer breweries much lower temperatures all round are carried than in ale breweries, and fermentation is carried on in rooms kept to a temperature of from 36 to 40° F. Before racking, and after settling, the beer is put into closed vessels, where a small amount of fermenting wort is added to it, the quantity being just sufficient to cause a slight fermentation of the whole of the beer without making it turbid. When in these closed vessels the beer is charged with more gas, and becomes wonderfully bright. It is then casked under a slight pressure, and the casks are at once shived, so that the pressure in them is retained. When the cask is removed to the rather warmer cellars of the consumer, the rise in temperature causes more gas to generate, the final result being that when the beer is drawn off it foams in the glass, but after a minute it becomes remarkably bright, clear, and sparkling—a condition which is much appreciated. It is, of course, necessary to keep lager beer brewed and racked under these conditions at a low temperature, as otherwise, owing to its low stability due to the absence of alcohol, it would be liable to the attacks of base ferments and would putrefy.

## The Design of Freezing Works.

**Some General Principles.**—The important elements of a well-designed freezing works may be usefully summarised as follows:—

(1) Facilities for easy transport to and from the works.

(2) The configuration of the ground on which the works are built.

(3) A plentiful water-supply.

There are many other points which have to be considered and weighed by an engineer before he finally decides on the site for the works, but these are the principal points which should govern his decision. Taking these points in numerical order, and considering them, it will be obvious that for transport purposes the works must be built on a line of railway, so that the railway trucks containing fat stock from up country can be run right into the works, and the stock unloaded right into the drafting yards adjoining the slaughter-house. Approach by road must also be provided, so that stock from the neighbouring country can be driven right into the works. The unloading platform and the road must also be in connection with the stock paddocks, so that on the arrival of a large herd some of the cattle can be turned out for a day or two until they are wanted.

In some cases the works can be placed near the sea or on a navigable river. In this case the problem of getting the prepared meat away is considerably simplified, and the carcasses can be run from the chilling or freezing chambers straight to the ship; more often, however, the carcasses have to be sent away by rail, and in this case the best arrangement is to have the slaughter-houses and freezing chambers between two sidings, so that the live stock comes in on one side of the works and the frozen carcasses go out on the other.

The second requirement is one which necessitates very careful consideration, as it determines the exact position of the buildings. If animals are to be killed and meat dressed and exported at the lowest possible cost, the buildings must be arranged so that a great deal of the necessary moving of the meat in its various stages can be done by gravity and not by machinery. This is a point which is frequently neglected when new works are being designed, and, as a consequence, when the business expands steam-engines have to be put down in all sorts of positions, and frequently yards away from the boilers which supply them, in order to work elevators and conveyors to carry the various products about the works. The guiding principle is frequently lost sight of, but it is an exceedingly simple one. When an article of commerce is bulky and of comparatively low intrinsic value, the difference between loss and profit in dealing with it depends on the observance of economies, which are quite immaterial when the article is of small bulk and of comparatively high intrinsic value. For example, when mutton only fetches 4d. per lb., a fraction of a penny per lb. in the works cost makes a vast difference, but with an article like copper, tin, or gold, which is of a high price per lb., a fraction of a penny in the works cost is of much less consequence. The site selected for the works should, if possible, be on the side of a hill; the cattle and sheep pens highest, then the crush, and then the slaughter-house; the offal and hide sheds should adjoin the killing-shed, and the gradient should be such that the dressed carcasses will travel by gravity to the chilling, freezing, and selecting rooms. After the carcasses are chilled or frozen, gravity comes in again, and by its aid they travel down to the railway loading-up platform or to the ship's side, as the case may be, ready for dispatch. If the site should be, as it often is, flat, these gradients can be artificially secured by an intelligent arrangement of the buildings, the primary elevation being obtained by making the stock walk to the slaughter-house, which is placed on an upper floor.

The principal profit of a freezing business should be made out of the by-products. Hoofs, horns, and bones can all be profitably disposed of. The hides, and the tallow from the surplus fat, are exceedingly valuable, and meat that is not good enough for export as dressed meat can be turned into extract with profitable results. Each part of the animal has its use and destination, and the works should be so planned that these various portions can reach that section of the building set apart for their treatment with the minimum of handling. By observing these principles some freezing works have, owing to their low works costs, been able to take sheep from the squatters and kill and freeze them for nothing, the whole of the works costs and profits coming out of the money obtained for the by-products. It is needless to point out what an advantage these works have over those run on less economical lines. If the works costs are low, also, more money can be devoted to improving the brand of the mutton or beef shipped, and anyone who knows anything of the business is aware what a good brand means.

The necessity of a good and plentiful water-supply is so obvious that it is hardly necessary to dwell upon this point. Scarce water means dirty works and a high cost for power and freezing, while a plentiful supply of water means the opposite.

The freezing and chilling houses are like ordinary cold storage buildings, and so long as the insulation is good and the entrances and exits well arranged, it does not much matter whether the shell of the building is built of wood or brick, the proper material being the cheapest in the particular locality. The air-blast or direct-expansion systems of freezing are both very suitable, the former, generally speaking, being cheapest in first cost and the latter in operating expenses. The whole of the boilers and engines, for whatever purpose they are required, should be placed close together in one power-house, and the various workshops, such as smithy, carpenter's shop, box-making shop, etc., where power is required, should be placed round it. Small steam-engines and long steam-pipes should be avoided, and if power is wanted at outlying parts of the yard, it should be transmitted electrically to motors placed where the power is required. In well-designed works there is no reason why there should be more than one battery of boilers or one chimney-shaft. Unless absolutely essential for some special reason, electric motors should not be used for driving the refrigerating machinery, the driving of an ammonia compressor, for instance, being about the most trying work an electric motor can be put to. For all other work motors are very suitable.

The arrangement of the conveyors and overhead tracks must be very carefully considered, and, if good results are to be obtained, as much care must be expended on them as on the laying out of a railway goods-yard. A good type of switch is necessary, and the crossings and points must be placed so that the trolleys can be easily switched on to the lines which will carry each brand of meat to its particular destination.

## *Refrigeration in the Dairy.*

**The Cooling of Cream.**—If cream is churned at a temperature of 48° F., not only can larger returns be obtained from a given quantity, but a much better class of butter can be produced than if the cream is churned at the higher temperatures that rule in summer. That there is a proper temperature for churning is, of course, housewife's knowledge, and in the old days churns used to be sold fitted with a thermometer and with jackets which could be filled with cold water in summer and warm water in winter. These were quite unsatisfactory, though perhaps better than nothing. In the modern creamery the milk is delivered by the farmers at a temperature of from about 65 to 70° F. in summer and 40 to 50° F. in winter. In quite a number of dairies, though unfortunately in not all, the milk is pasteurized by being heated up in a special apparatus to a temperature of about 180° F., the object being to destroy the bacteria, which are so active, and which find such a congenial home in milk. This temperature is not sufficiently high to afford complete immunity from bacteria, but practically it is sufficient, and with it there is no danger of burning.

With an Alpha Laval separator separation will be complete at any temperature from 100° upwards, but generally it has been found that the cream will more readily separate from the milk at a temperature of 160° than at any other. Though they have nothing to do with each other, the high temperature which is necessary for pasteurization is desirable for separating, and two objects are thus attained by the one heating. After pasteurization and separation the milk is cooled down by being passed over a cooler of the Baudelot type—very similar to those used in breweries for cooling wort—through which well water is kept running, and is returned to the farmers for feeding calves, etc. The cream is cooled rapidly by being passed over a circular capillary cooler, through the corrugations of which well water is circulating. This reduces its temperature to about 70° F., which is about as low as it is possible to get it with ordinary well water, though in parts of Ireland, notably County Kerry, the temperature of the springs is much lower.

The more rapidly milk is cooled, the better for it and the longer it can be preserved. It is, therefore, desirable to still further reduce the temperature of the cream to about 50°. In view of the fact that the subsequent ripening again raises the temperature of the cream to 70°, this cooling might appear to be unnecessary, but it is not so, as any tendency to bacterial activity and the absorption of aerial ferments is counteracted by it. At this temperature, then, the cream is placed in vats, where it ripens. Practically it ferments, and, due to this lactic fermentation, the temperature, as stated above, rises. In from fifteen to twenty hours the cream is ready to be churned, but, in order that the fat globules may not be broken and the butter become oily, and in order also that the yield may be as large as possible, the cream is again cooled by being run over a capillary cooler on its way from the ripening vat to the churn, which it reaches at a temperature of 48°. In some forty-five minutes churning is complete and the butter floats on the top of the whey in a fine firm condition.

This, then, is the process. The necessity for refrigeration will be seen in the low temperature which it is absolutely necessary to keep if it is wished to obtain the best results. It is desirable also to chill the butter and the milk before it is sent to market.

Fig. 112[1] illustrates a small refrigerating machine and plant for cooling cream by means of a circular capillary cooler. The cream is run on to the distributing saucer from the ripening vats, and flows gently and evenly down over the wide flutings of the cooler into the collecting saucer at the bottom and away to the churn. Brine chilled in the evaporator of the refrigerating machine is pumped round through the flutings of the cooler and absorbs heat from the cream. The distribution of the cream over the cooler must be regulated so as to ensure its all being of an even temperature.

In some dairies the ripening vats are placed so as to command the churns, the cream being run from them over the cooler and into the churns by gravity. Where this is done, the regulation of the flow over the coolers is not difficult; but where for some reason or other the cream has to be pumped, care should be taken to regulate the suction of the pump to give an even flow.

The essence of a successful dairy is cleanliness, and great care should be taken to arrange that every vessel in

[1] Messrs William Douglas & Sons, Ltd., Putney, London. A somewhat similar block has appeared in *Ice and Cold Storage*.

contact with the milk or cream can be thoroughly scalded down each time after it has been used.

**Milk-cooling and Preserving.**—Undoubtedly the surest precaution against disease in milk is pasteurization, but, though common in Denmark and in some of the Irish dairies where milk is heated up to from 200 to 212° F.,

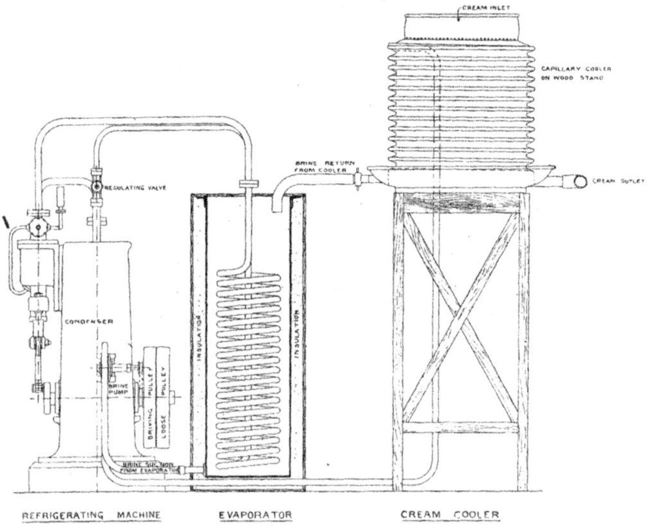


Fig. 112.

this country has not yet advanced so far. The next best thing to pasteurizing the domestic milk supply is to cool it to a low temperature before it leaves the dairy. This is done by running it over a vertical milk-cooler constructed of fluted and tinned copper. The milk runs down the outside of the cooler in a thin stream and is cooled by refrigerated water or brine, which circulates inside the coils. The apparatus, in fact, is very much the same as that used in breweries for cooling wort.

Pasteurization, which only destroys full-grown bacteria, must not be confused with sterilization. The latter process consists of heating the milk up to some 200° F. to destroy the bacteria, cooling it down, and letting it stand until the bacteria spores which have not been destroyed have had time to develop. In order to destroy these spores, the process is then repeated. It is essential that during and after sterilization the milk should be protected from contact with the outside air. It is accordingly bottled in hermetically sealed bottles and offered for sale. After sterilization the milk will keep for any length of time, so long as it is not exposed to the atmosphere and the aerial ferments it contains.

### THE CRYSTALLIZATION OF PARAFFIN.

During the process of distilling the natural oil shales, a number of oils of different specific gravities are obtained. With the more volatile kinds refrigeration is not concerned, but it is used extensively in assisting the crystallization of those which are solid at ordinary temperatures and which, when mixed with stearine, are extensively used for making candles. During the cold winter months no great difficulty is found in separating the heavy paraffin from the other oils, as it will crystallize out; but in the summer months the temperature is too high, and crystallization will not take place to its full extent. One of the first applications of refrigeration was to cooling paraffin in the summer, and a large saving was made by Kirk's machine, which was used for cooling oil in some works near Bathgate. Until comparatively recently the cooling of the oil and the crystallization of the paraffin was brought about by revolving a drum filled with refrigerated brine in a tank containing the oil. The paraffin coming in contact with the cold metal congealed on its surface, and was scraped off the drum as it revolved. This process, though it ensured the rapid cooling of a large quantity of oil, was not satisfactory for the following reasons:—

The object of cooling paraffin—a process essentially of crystallization and filtration—is to separate efficiently the paraffin from the oil. This can only be done by allowing the paraffin crystals to grow to a considerable size, so that the oil will drain away from them. Owing, however, to the high average viscosity of the oil and the close interlacing of the crystals, about 75 per cent. of the oil cannot be separated by draining alone, and filtration has to be resorted to. For this purpose the solidified cakes of paraffin have to be broken up, and the oil contained between and round the crystals liberated. The paraffin, which still contains its crystalline formation, is then passed through a filter-press, which separates out the balance of the oil. The rapid cooling on a drum did not give the crystals time to develop, and consequently the paraffin was only very faintly crystalline, and, therefore, in the best possible position for holding a large quantity of oil in its bulk.

A number of experiments were made with the object of cooling the paraffin more slowly and by artificial means, but, owing to the very low conductivity of paraffin, the time and space required was found to be too great. About 1884 Mr Beilby started some experiments at Oakbank Oil Works, and economized space by having narrow deep coolers standing on their sides, instead, as had been previously done, of having shallow flat coolers resting on

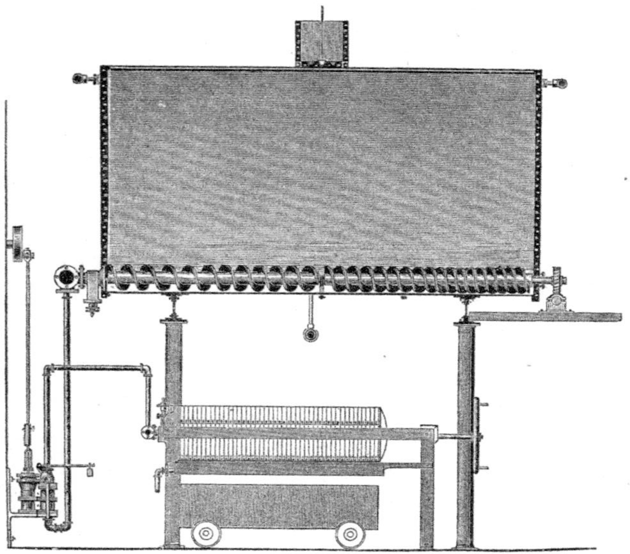

Fig 113.

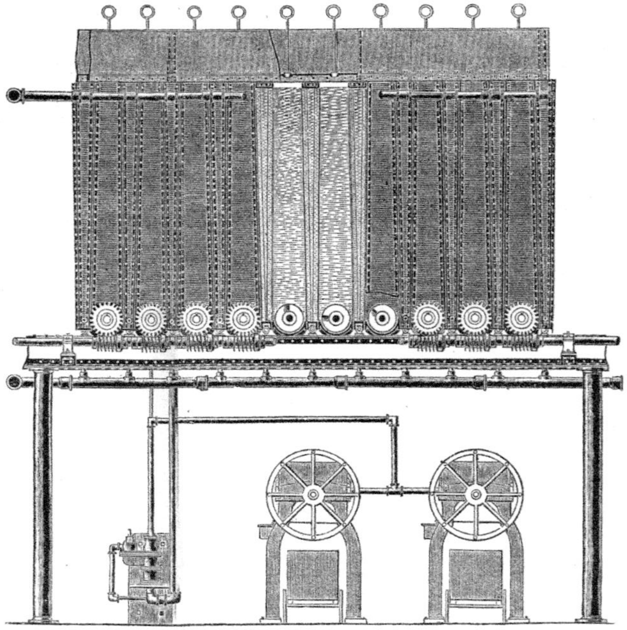

Fig. 114.

*To face page 391.*

their bottoms. It was found that a cooling cell 16 feet long, 8 feet 6 inches deep, and 12 inches wide, filled with paraffin solution at 100° F. and subjected to a temperature of 24°, would solidify in from 70 to 75 hours, with very satisfactory results as regards the size of the crystals. The complete apparatus designed by Mr Beilby is shown in figs. 113 and 114[1]; the crushing of the cakes of paraffin after solidification is done by the screws in the bottom of each cell. These are rotated by means of the shaft and worm-wheel shown, and the crushed paraffin is delivered to the filter-presses shown below. In section the cells will be seen to be narrower at the bottom, or exactly opposite to that of an ice-can. This is done so that when the screw revolves and cuts away the base of the solid block, the taper causes the block to settle down on the screw, which gradually shears it away and delivers it to the presses. Owing to the solid state of the mass of paraffin, and the difficulty experienced in getting it from the cells to the filter-presses, the closed pipes shown in the figures had to be abandoned and replaced by open wooden troughs. The plant which was worked in connection with a Pontifex and Wood absorption machine gave very good results at a fairly moderate cost.

In another system the mother oil is placed in a cold room in shallow trays which are jacketed with refrigerated brine; the paraffin solidifies, and the advantage claimed is that the paraffin of various specific gravities can be separated from each other.

## Chocolate-Cooling.

Chocolate to be satisfactory must be cooled quickly, and

[1] *Journal of the Society of Chemical Industry*, 1885.

for this reason, before the days of mechanical refrigeration, chocolate was only made in the winter when the naturally low temperature would allow it to set and harden. Refrigeration has changed this, and now, after manufacture, the block chocolate and covered goods are placed in a room which is artificially kept at a temperature of about 45° F. In this temperature the chocolate sets hard, and can readily be removed from the moulds. Chocolate-covered goods present greater difficulties; the centres are dipped in hot chocolate which adheres to them, they are then removed to the chill room and cooled. The covering-room, however, has to be kept at a comfortable temperature, as a number of people are employed in it; 65° is the highest possible, as above this temperature the chocolate will not set at all. The covering-tables have to be kept hot and the chocolate in a liquid state by steam-coils or by gas, and at the same time the room has to be kept cool and well ventilated by a circulation of air. In some cases the covering-room is cooled by brine-pipes above a false ceiling; in others by a gentle circulation of cooled air. The chilling-room is furnished with shelves, upon which the trays are placed, and can be cooled by any of the recognized systems, the most important point being the dryness of the atmosphere, so that there can be no chance of moisture being deposited on the goods being cooled. The temperature in the chill chamber should never exceed 55° F., or the chocolate will not be cooled sufficiently rapidly, and will go off colour and generally deteriorate. From particulars taken during manufacture it has been found that when the chocolate is moulded the temperature is 83° F., and the highest temperature at which it will leave the moulds is 58° F. There are several methods of cooling chocolate, and manufacturers differ considerably as to the proper temperatures. Fig. 115[1] illustrates the thermo-gradient chocolate-cooler, which is worked in conjunction with a cold dry-air machine. The very cold air from the machine at a temperature below zero enters the chamber at the bottom, and, as it absorbs heat from the chocolate, gradually rises to the top, the volume of air being so arranged that any required temperature—45 or 50° F.—can be maintained at this point. The trays con-

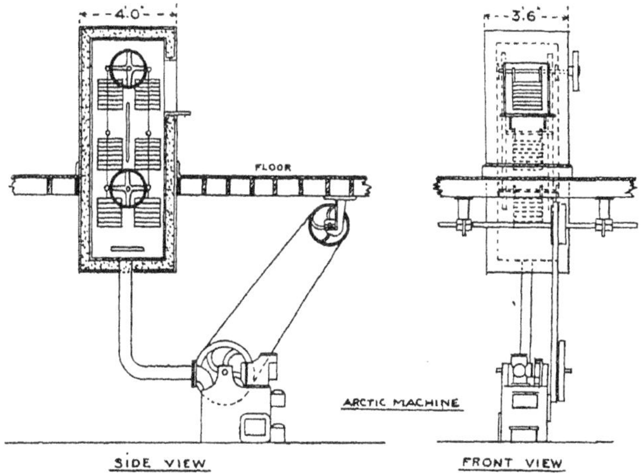


Fig. 115.

taining the chocolate are put on to the endless chain, which is kept slowly revolving, and gradually descend from the warmer air at the top to the coldest point at the bottom. Passing through this it gradually ascends into the warmer air, and is removed from the chamber at a temperature sufficiently high to prevent the moisture present in the outside air depositing upon it.

[1] Messrs T. & W. Cole, Ltd., London.

A rotary chocolate-cooler[1] possessing some novel features has recently been introduced. This consists of a circular chamber containing shelves arranged in compartments which are fixed to a central vertical spindle rotated by a hand-wheel fixed outside the cooler. The compartments are all numbered, and, as each arrives opposite the door, its number is automatically indicated on a dial placed above the hand-wheel. The air inside the chamber is cooled by brine-drums, and air circulation is maintained by a small fan.

[1] The Express Rotary Cooler Syndicate, Ltd., London.

## PROPERTIES OF AMMONIA.

PROF. DE V. WOOD. *Extended by* G. DAVIDSON.

| Temperature. | | Pressure, Absolute. | | Gauge pressure, lb. per square inch. | Heat of vaporization, thermal units. | Volume of vapour per lb., cubic feet. | Volume of liquid per lb., cubic feet. | Weight of vapour in lbs. per cubic foot. | Weight of liquid in lbs. per cubic foot. | Temperature. |
|---|---|---|---|---|---|---|---|---|---|---|
| Degrees F. | Absolute. | Lbs. per square foot. P. | Lbs. per square inch. p. | | | | | | | Degrees F. |
| t | T | P | p | p′ | L | V | v | W | w | t |
| −40 | 420·66 | 1539·90 | 10·69 | −4·01 | 579·67 | 24·388 | ·02348 | ·0410 | 42·589 | −40 |
| 39 | 1 | 1584·43 | 11·00 | −3·7 | 579·07 | 23·735 | ·02351 | ·0421 | 42·535 | 39 |
| 38 | 2 | 1630·03 | 11·32 | −3·38 | 578·47 | 23·102 | ·02354 | ·0433 | 42·483 | 38 |
| 37 | 3 | 1676·71 | 11·64 | −3·06 | 577·88 | 22·488 | ·02357 | ·0444 | 42·427 | 37 |
| 36 | 4 | 1724·51 | 11·98 | −2·72 | 577·27 | 21·895 | ·02359 | ·0457 | 42·391 | 36 |
| −35 | 425·66 | 1773·43 | 12·31 | −2·39 | 576·68 | 21·321 | ·02362 | ·0469 | 42·337 | −35 |
| 34 | 6 | 1823·50 | 12·66 | −2·04 | 576·08 | 20·763 | ·02364 | ·0482 | 42·301 | 34 |
| 33 | 7 | 1874·73 | 13·02 | −1·68 | 575·48 | 20·221 | ·02366 | ·0495 | 42·265 | 33 |
| 32 | 8 | 1927·17 | 13·38 | −1·32 | 574·89 | 19·708 | ·02368 | ·0507 | 42·213 | 32 |
| 31 | 9 | 1980·78 | 13·75 | −0·95 | 574·39 | 19·204 | ·02371 | ·0521 | 42·176 | 31 |
| −30 | 430·66 | 2035·69 | 14·13 | −0·57 | 573·69 | 18·693 | ·02374 | ·0535 | 42·123 | −30 |
| 29 | 1 | 2091·83 | 14·53 | −0·17 | 573·08 | 18·225 | ·02378 | ·0549 | 42·052 | 29 |
| 28 | 2 | 2149·23 | 14·92 | +0·22 | 572·48 | 17·759 | ·02381 | ·0563 | 42·000 | 28 |
| 27 | 3 | 2207·94 | 15·33 | +0·63 | 571·89 | 17·307 | ·02384 | ·0577 | 41·946 | 27 |
| 26 | 4 | 2267·97 | 15·75 | +1·05 | 571·28 | 16·869 | ·02387 | ·0593 | 41·893 | 26 |
| −25 | 435·66 | 2329·34 | 16·17 | +1·47 | 570·68 | 16·446 | ·02389 | ·0608 | 41·858 | −25 |
| 24 | 6 | 2392·09 | 16·61 | 1·91 | 570·08 | 16·034 | ·02392 | ·0624 | 41·806 | 24 |
| 23 | 7 | 2456·23 | 17·05 | 2·35 | 569·48 | 15·633 | ·02395 | ·0640 | 41·754 | 23 |
| 22 | 8 | 2520·45 | 17·50 | 2·8 | 568·88 | 15·252 | ·02398 | ·0656 | 41·701 | 22 |
| 21 | 9 | 2588·77 | 17·97 | 3·27 | 568·27 | 14·875 | ·02401 | ·0672 | 41·649 | 21 |
| −20 | 440·66 | 2657·23 | 18·45 | +3·75 | 567·67 | 14·507 | ·02403 | ·0689 | 41·615 | −20 |
| 19 | 1 | 2727·17 | 18·94 | 4·24 | 567·06 | 14·153 | ·02406 | ·0706 | 41·563 | 19 |
| 18 | 2 | 2798·62 | 19·43 | 4·73 | 566·43 | 13·807 | ·02409 | ·0725 | 41·511 | 18 |
| 17 | 3 | 2871·61 | 19·94 | 5·24 | 565·85 | 13·475 | ·02411 | ·0742 | 41·480 | 17 |
| 16 | 4 | 2946·17 | 20·46 | 5·76 | 565·25 | 13·150 | ·02414 | ·0760 | 41·425 | 16 |
| −15 | 445·66 | 3022·31 | 20·99 | +6·29 | 564·64 | 12·834 | ·02417 | ·0779 | 41·374 | −15 |
| 14 | 6 | 3100·07 | 21·53 | 6·83 | 564·04 | 12·527 | ·02420 | ·0798 | 41·322 | 14 |
| 13 | 7 | 3179·45 | 22·08 | 7·38 | 563·43 | 12·230 | ·02423 | ·0818 | 41·271 | 13 |
| 12 | 8 | 3260·52 | 22·64 | 7·94 | 562·82 | 11·939 | ·02425 | ·0838 | 41·237 | 12 |
| 11 | 9 | 3343·29 | 23·22 | 8·52 | 562·21 | 11·659 | ·02428 | ·0858 | 41·186 | 11 |

## PROPERTIES OF AMMONIA—(*continued*).

| Temperature. | | Pressure, Absolute. | | Gauge pressure, lb. per square inch. | Heat of vaporization, thermal units. | Volume of vapour per lb., cubic feet. | Volume of liquid per lb., cubic feet. | Weight of vapour in lbs. per cubic foot. | Weight of liquid in lbs. per cubic foot. | Temperature. |
|---|---|---|---|---|---|---|---|---|---|---|
| Degrees F. | Absolute. | Lbs. per square foot. P. | Lbs. per square inch. p. | | | | | | | Degrees F. |
| t | T | P | p | p' | L | V | v | W | w | t |
| −10 | 450·66 | 3427·75 | 23·80 | +9·10 | 561·61 | 11·385 | ·02431 | ·0878 | 41·135 | −10 |
| 9 | 1 | 3513·97 | 24·40 | 9·70 | 560·99 | 11·117 | ·02434 | ·0899 | 41·084 | 9 |
| 8 | 2 | 3601·97 | 25·01 | 10·31 | 560·39 | 10·860 | ·02437 | ·0921 | 41·034 | 8 |
| 7 | 3 | 3691·75 | 25·64 | 10·94 | 559·78 | 10·604 | ·02439 | ·0943 | 41·000 | 7 |
| 6 | 4 | 3783·37 | 26·27 | 11·57 | 559·17 | 10·362 | ·02442 | ·0965 | 40·950 | 6 |
| −5 | 455·66 | 3876·85 | 26·92 | +12·22 | 558·56 | 10·125 | ·02445 | ·0988 | 40·900 | −5 |
| 4 | 6 | 3972·62 | 27·59 | 12·89 | 557·94 | 9·894 | ·02448 | ·1011 | 40·849 | 4 |
| 3 | 7 | 4069·48 | 28·26 | 13·56 | 557·33 | 9·669 | ·02451 | ·1034 | 40·799 | 3 |
| 2 | 8 | 4168·70 | 28·95 | 14·25 | 556·73 | 9·449 | ·02454 | ·1058 | 40·749 | 2 |
| 1 | 9 | 4269·90 | 29·65 | 14·95 | 556·11 | 9·234 | ·02457 | ·1083 | 40·700 | 1 |
| 0 | 460·66 | 4373·10 | 30·37 | +15·67 | 555·50 | 9·028 | ·02461 | ·1107 | 40·650 | 0 |
| +1 | 1 | 4478·32 | 31·10 | 16·40 | 554·80 | 8·825 | ·02463 | ·1133 | 40·601 | +1 |
| 2 | 2 | 4585·60 | 31·84 | 17·14 | 554·27 | 8·630 | ·02466 | ·1159 | 40·551 | 2 |
| 3 | 3 | 4694·96 | 32·60 | 17·90 | 553·65 | 8·436 | ·02469 | ·1186 | 40·502 | 3 |
| 4 | 4 | 4806·46 | 33·38 | 18·68 | 553·04 | 8·250 | ·02472 | ·1212 | 40·453 | 4 |
| +5 | 465·66 | 4920·11 | 34·16 | +19·46 | 552·43 | 8·070 | ·02475 | ·1240 | 40·40[illegible] | +5 |
| 6 | 6 | 5035·95 | 34·97 | 20·27 | 551·81 | 7·892 | ·02478 | ·1267 | 40·355 | 6 |
| 7 | 7 | 5153·99 | 35·79 | 21·09 | 551·19 | 7·717 | ·02480 | ·1296 | 40·322 | 7 |
| 8 | 8 | 5274·28 | 36·63 | 21·93 | 550·58 | 7·553 | ·02483 | ·1324 | 40·274 | 8 |
| 9 | 9 | 5396·83 | 37·48 | 22·78 | 549·96 | 7·388 | ·02486 | ·1353 | 40·225 | 9 |
| +10 | 470·66 | 5521·71 | 38·34 | +23·64 | 549·35 | 7·229 | ·02490 | ·1383 | 40·160 | +10 |
| 11 | 1 | 5649·48 | 39·23 | 24·53 | 548·73 | 7·075 | ·02493 | ·1413 | 40·112 | 11 |
| 12 | 2 | 5778·50 | 40·13 | 25·43 | 548·11 | 6·924 | ·02496 | ·1444 | 40·064 | 12 |
| 13 | 3 | 5910·52 | 41·04 | 26·34 | 547·49 | 6·786 | ·02499 | ·1474 | 40·016 | 13 |
| 14 | 4 | 6044·96 | 41·98 | 27·28 | 546·88 | 6·632 | ·02502 | ·1508 | 39·968 | 14 |
| +15 | 475·66 | 6182·00 | 42·94 | +28·24 | 546·26 | 6·491 | ·02505 | ·1541 | 39·920 | +15 |
| 16 | 6 | 6321·24 | 43·90 | 29·20 | 545·63 | 6·355 | ·02508 | ·1573 | 39·872 | 16 |
| 17 | 7 | 6463·24 | 44·88 | 30·18 | 545·01 | 6·222 | ·02511 | ·1607 | 39·824 | 17 |
| 18 | 8 | 6607·77 | 45·89 | 31·19 | 544·39 | 6·093 | ·02514 | ·1641 | 39·777 | 18 |
| 19 | 9 | 6754·90 | 46·91 | 32·21 | 543·77 | 5·966 | ·02517 | ·1676 | 39·729 | 19 |

## PROPERTIES OF AMMONIA—(*continued*).

| Temperature. | | Pressure, Absolute. | | Gauge pressure, lb. per square inch. | Heat of vaporization, thermal units. | Volume of vapour per lb., cubic feet. | Volume of liquid per lb., cubic feet. | Weight of vapour in lbs. per cubic foot. | Weight of liquid in lbs. per cubic foot. | Temperature. |
|---|---|---|---|---|---|---|---|---|---|---|
| Degrees F. | Absolute. | Lbs. per square foot. *P*. | Lbs. per square inch. *p*. | | | | | | | Degrees F. |
| *t* | *T* | *P* | *p* | *p′* | *L* | *V* | *v* | *W* | *w* | *t* |
| +20 | 480·66 | 6904·68 | 47·95 | +33·25 | 543·15 | 5·843 | ·02520 | ·1711 | 39·682 | +20 |
| 21 | 1 | 7057·15 | 49·01 | 34·31 | 542·53 | 5·722 | ·02523 | ·1748 | 39·635 | 21 |
| 22 | 2 | 7212·33 | 50·09 | 35·39 | 541·90 | 5·605 | ·02527 | ·1784 | 39·572 | 22 |
| 23 | 3 | 7370·27 | 51·18 | 36·48 | 541·28 | 5·488 | ·02529 | ·1822 | 39·541 | 23 |
| 24 | 4 | 7530·96 | 52·30 | 37·60 | 540·66 | 5·378 | ·02533 | ·1860 | 39·479 | 24 |
| +25 | 485·66 | 7694·52 | 53·43 | +38·73 | 540·03 | 5·270 | ·02536 | ·1897 | 39·432 | +25 |
| 26 | 6 | 7860·89 | 54·59 | 39·89 | 539·41 | 5·163 | ·02539 | ·1937 | 39·386 | 26 |
| 27 | 7 | 8030·16 | 55·76 | 41·06 | 538·78 | 5·058 | ·02542 | ·1977 | 39·339 | 27 |
| 28 | 8 | 8202·38 | 50·96 | 42·26 | 538·16 | 4·960 | ·02545 | ·2016 | 39·292 | 28 |
| 29 | 9 | 8377·56 | 58·17 | 43·47 | 537·53 | 4·858 | ·02548 | ·2059 | 39·246 | 29 |
| +30 | 490·66 | 8555·74 | 59·42 | +44·72 | 536·91 | 4·763 | ·02551 | ·2099 | 39·200 | +30 |
| 31 | 1 | 8736·96 | 60·67 | 45·97 | 536·28 | 4·668 | ·02554 | ·2142 | 39·115 | 31 |
| 32 | 2 | 8921·26 | 61·95 | 47·25 | 535·66 | 4·577 | ·02557 | ·2185 | 39·108 | 32 |
| 33 | 3 | 9108·71 | 63·25 | 48·55 | 535·03 | 4·486 | ·02561 | ·2229 | 39·047 | 33 |
| 34 | 4 | 9299·32 | 64·58 | 49·88 | 534·40 | 4·400 | ·02564 | ·2273 | 39·001 | 34 |
| +35 | 495·66 | 9493·07 | 65·92 | +51·22 | 533·78 | 4·314 | ·02568 | ·2318 | 38·940 | +35 |
| 36 | 6 | 9690·04 | 67·29 | 52·59 | 533·13 | 4·234 | ·02571 | ·2362 | 38·894 | 36 |
| 37 | 7 | 9890·75 | 68·68 | 53·98 | 532·52 | 4·147 | ·02574 | ·2413 | 38·850 | 37 |
| 38 | 8 | 10093·91 | 70·09 | 55·39 | 531·89 | 4·068 | ·02578 | ·2458 | 38·789 | 38 |
| 39 | 9 | 10300·88 | 71·53 | 56·83 | 531·26 | 3·989 | ·02582 | ·2507 | 38·721 | 39 |
| +40 | 500·66 | 10511·16 | 72·99 | +58·29 | 530·63 | 3·915 | ·02585 | ·2554 | 38·684 | +40 |
| 41 | 1 | 10724·95 | 74·48 | 59·78 | 529·99 | 3·839 | ·02588 | ·2605 | 38·639 | 41 |
| 42 | 2 | 10942·18 | 75·99 | 61·29 | 529·36 | 3·766 | ·02591 | ·2655 | 38·595 | 42 |
| 43 | 3 | 11162·93 | 77·52 | 62·82 | 528·73 | 3·695 | ·02594 | ·2706 | 38·550 | 43 |
| 44 | 4 | 11387·21 | 79·08 | 64·38 | 528·10 | 3·627 | ·02597 | ·2757 | 38·499 | 44 |
| +45 | 505·66 | 11615·12 | 80·66 | +65·96 | 527·47 | 3·559 | ·02600 | ·2809 | 38·461 | +45 |
| 46 | 6 | 11846·64 | 82·27 | 67·57 | 526·83 | 3·493 | ·02603 | ·2863 | 38·417 | 46 |
| 47 | 7 | 12081·80 | 83·90 | 69·20 | 526·20 | 3·428 | ·02606 | ·2917 | 38·373 | 47 |
| 48 | 8 | 12320·71 | 85·56 | 70·86 | 525·57 | 3·362 | ·02609 | ·2974 | 38·328 | 48 |
| 49 | 9 | 12563·36 | 87·25 | 72·55 | 524·93 | 3·303 | ·02612 | ·3027 | 38·284 | 49 |

## PROPERTIES OF AMMONIA—(*continued*).

| Temperature. | | Pressure, Absolute. | | Gauge pressure, lb. per square inch. | Heat of vaporization, thermal units. | Volume of vapour per lb., cubic feet. | Volume of liquid per lb., cubic feet. | Weight of vapour in lbs. per cubic foot. | Weight of liquid in lbs. per cubic foot. | Temperature. |
|---|---|---|---|---|---|---|---|---|---|---|
| Degrees F. | Absolute. | Lbs. per square foot. P. | Lbs. per square inch. p. | | | | | | | Degrees F. |
| *t* | *T* | *P* | *p* | *p'* | *L* | *V* | *v* | *W* | *w* | *t* |
| +50 | 510·66 | 12809·91 | 88·96 | + 74·26 | 524·30 | 3·242 | ·02616 | ·3084 | 38·226 | +50 |
| 51 | 1 | 13060·21 | 90·70 | 76·00 | 523·66 | 3·182 | ·02620 | ·3143 | 38·167 | 51 |
| 52 | 2 | 13314·43 | 92·46 | 77·76 | 523·03 | 3·124 | ·02623 | ·3201 | 38·124 | 52 |
| 53 | 3 | 13572·52 | 94·25 | 79·55 | 522·39 | 3·069 | ·02626 | ·3258 | 38·080 | 53 |
| 54 | 4 | 13834·64 | 96·07 | 81·37 | 521·76 | 3·012 | ·02629 | ·3320 | 38·037 | 54 |
| +55 | 515·66 | 14100·74 | 97·92 | + 83·22 | 521·12 | 2·958 | ·02632 | ·3380 | 37·994 | +55 |
| 56 | 6 | 14370·92 | 99·80 | 85·10 | 520·48 | 2·905 | ·02636 | ·3442 | 37·936 | 56 |
| 57 | 7 | 14645·18 | 101·70 | 87·00 | 519·84 | 2·853 | ·02639 | ·3505 | 37·893 | 57 |
| 58 | 8 | 14923·98 | 103·64 | 88·94 | 519·20 | 2·802 | ·02643 | ·3568 | 37·835 | 58 |
| 59 | 9 | 15206·28 | 105·60 | 90·90 | 518·57 | 2·753 | ·02646 | ·3632 | 37·793 | 59 |
| +60 | 520·66 | 15493·09 | 107·59 | + 92·89 | 517·93 | 2·705 | ·02651 | ·3697 | 37·736 | +60 |
| 61 | 1 | 15784·23 | 109·61 | 94·91 | 517·29 | 2·658 | ·02654 | ·3762 | 37·678 | 61 |
| 62 | 2 | 16079·67 | 111·66 | 96·96 | 516·65 | 2·610 | ·02658 | ·3831 | 37·622 | 62 |
| 63 | 3 | 16379·51 | 113·75 | 99·05 | 516·01 | 2·565 | ·02661 | ·3898 | 37·579 | 63 |
| 64 | 4 | 16683·75 | 115·86 | 101·16 | 515·37 | 2·520 | ·02665 | ·3968 | 37·523 | 64 |
| +65 | 525·66 | 16992·50 | 118·03 | +103·33 | 514·73 | 2·476 | ·02668 | ·4039 | 37·481 | +65 |
| 66 | 6 | 17305·70 | 120·18 | 105·48 | 514·09 | 2·433 | ·02671 | ·4110 | 37·439 | 66 |
| 67 | 7 | 17623·45 | 122·38 | 107·68 | 513·45 | 2·389 | ·02675 | ·4189 | 37·383 | 67 |
| 68 | 8 | 17945·89 | 124·62 | 109·92 | 512·81 | 2·351 | ·02678 | ·4254 | 37·341 | 68 |
| 69 | 9 | 18272·81 | 126·89 | 112·19 | 512·16 | 2·310 | ·02682 | ·4329 | 37·285 | 69 |
| +70 | 530·66 | 18604·53 | 129·19 | +114·49 | 511·52 | 2·272 | ·02686 | ·4401 | 37·230 | +70 |
| 71 | 1 | 18941·00 | 131·54 | 116·84 | 510·87 | 2·233 | ·02689 | ·4479 | 37·188 | 71 |
| 72 | 2 | 19282·21 | 133·90 | 119·20 | 510·22 | 2·194 | ·02693 | ·4558 | 37·133 | 72 |
| 73 | 3 | 19628·32 | 136·31 | 121·61 | 509·58 | 2·153 | ·02697 | ·4645 | 37·079 | 73 |
| 74 | 4 | 19979·22 | 138·74 | 124·04 | 508·93 | 2·122 | ·02700 | ·4712 | 37·037 | 74 |
| +75 | 535·66 | 20335·16 | 141·22 | +126·52 | 508·29 | 2·087 | ·02703 | ·4791 | 36·995 | +75 |
| 76 | 6 | 20696·00 | 143·72 | 129·02 | 507·64 | 2·052 | ·02706 | ·4873 | 36·954 | 76 |
| 77 | 7 | 21061·85 | 146·26 | 131·56 | 506·99 | 2·017 | ·02710 | ·4957 | 36·900 | 77 |
| 78 | 8 | 21432·82 | 148·84 | 134·14 | 506·34 | 1·995 | ·02714 | ·5012 | 36·845 | 78 |
| 79 | 9 | 21808·85 | 151·45 | 136·75 | 505·69 | 1·952 | ·02717 | ·5123 | 36·805 | 79 |

## Properties of Ammonia—(*continued*).

| Temperature. | | Pressure, Absolute. | | | | | | | | Temperature. |
|---|---|---|---|---|---|---|---|---|---|---|
| Degrees F. | Absolute. | Lbs. per square foot. P. | Lbs. per square inch. p. | Gauge pressure, lb. per square inch. | Heat of vaporization, thermal units. | Volume of vapour per lb., cubic feet. | Volume of liquid per lb., cubic feet. | Weight of vapour in lbs. per cubic foot. | Weight of liquid in lbs. per cubic foot. | Degrees F. |
| *t* | *T* | *P* | *p* | *p'* | *L* | *V* | *v* | *W* | *w* | *t* |
| +80 | 540·66 | 22190·15 | 154·10 | +139·40 | 505·05 | 1·921 | ·02721 | ·5205 | 36·751 | + 80 |
| 81 | 1 | 22576·51 | 156·78 | 142·08 | 504·40 | 1·889 | ·02725 | ·5294 | 36·696 | 81 |
| 82 | 2 | 22968·88 | 159·50 | 144·80 | 503·75 | 1·858 | ·02728 | ·5382 | 36·657 | 82 |
| 83 | 3 | 23365·38 | 162·26 | 147·56 | 503·10 | 1·827 | ·02732 | ·5473 | 36·603 | 83 |
| 84 | 4 | 23767·81 | 165·05 | 150·35 | 502·45 | 1·799 | ·02736 | ·5558 | 36·549 | 84 |
| +85 | 545·66 | 24175·61 | 167·88 | +153·18 | 501·81 | 1·770 | ·02739 | ·5649 | 36·509 | + 85 |
| 86 | 6 | 24588·92 | 170·75 | 156·05 | 501·15 | 1·741 | ·02743 | ·5744 | 36·456 | 86 |
| 87 | 7 | 25007·80 | 173·66 | 158·96 | 500·50 | 1·714 | ·02747 | ·5834 | 36·407 | 87 |
| 88 | 8 | 25432·16 | 176·61 | 161·91 | 499·85 | 1·687 | ·02751 | ·5927 | 36·350 | 88 |
| 89 | 9 | 25862·14 | 179·59 | 164·89 | 499·20 | 1·660 | ·02754 | ·6024 | 36·311 | 89 |
| +90 | 550·66 | 26297·88 | 182·62 | +167·92 | 498·55 | 1·634 | ·02758 | ·6120 | 36·258 | + 90 |
| 91 | 1 | 26739·88 | 185·69 | 170·99 | 497·89 | 1·608 | ·02761 | ·6219 | 36·219 | 91 |
| 92 | 2 | 27186·56 | 188·79 | 174·09 | 497·24 | 1·583 | ·02765 | ·6317 | 36·166 | 92 |
| 93 | 3 | 27639·43 | 191·94 | 177·24 | 496·59 | 1·558 | ·02769 | ·6418 | 36·114 | 93 |
| 94 | 4 | 28098·26 | 195·13 | 180·43 | 495·94 | 1·534 | ·02772 | ·6518 | 36·075 | 94 |
| +95 | 555·66 | 28563·00 | 198·35 | +183·65 | 495·29 | 1·510 | ·02776 | ·6622 | 36·023 | + 95 |
| 96 | 6 | 29033·86 | 201·62 | 186·92 | 494·63 | 1·486 | ·02780 | ·6729 | 35·971 | 96 |
| 97 | 7 | 29510·69 | 204·94 | 190·24 | 493·97 | 1·463 | ·02784 | ·6835 | 35·919 | 97 |
| 98 | 8 | 29993·52 | 208·29 | 193·59 | 493·32 | 1·442 | ·02787 | ·6934 | 35·881 | 98 |
| 99 | 9 | 30482·52 | 211·68 | 196·98 | 492·66 | 1·419 | ·02791 | ·7047 | 35·829 | 99 |
| + 00 | 560·66 | 30977·78 | 215·12 | +200·42 | 492·01 | 1·398 | ·02795 | ·7153 | 35·778 | +100 |

# INDEX.

PRINTED BY NEILL AND CO., LTD., EDINBURGH.

Printed in Dunstable, United Kingdom

91220852R00261